WATER RESOURCE PLANNING, DEVELOPMENT AND MANAGEMENT

DRINKING WATER

CHICAGO, PENNSYLVANIA, CONNECTICUT, MICHIGAN AND TRIBAL LANDS

WATER RESOURCE PLANNING, DEVELOPMENT AND MANAGEMENT

Additional books and e-books in this series can be found on Nova's website under the Series tab.

WATER RESOURCE PLANNING, DEVELOPMENT AND MANAGEMENT

DRINKING WATER

CHICAGO, PENNSYLVANIA, CONNECTICUT, MICHIGAN AND TRIBAL LANDS

ERWIN JON OROSZ
EDITOR

snova
New York

Library of Congress Cataloging-in-Publication Data

ISBN: 978-1-53617-284-3

Published by Nova Science Publishers, Inc. † New York

CONTENTS

PREFACE

The U.S. Geological Survey (USGS) Environmental Health Mission Area (EHMA) is providing comprehensive science on sources, movement, and transformation of contaminants and pathogens in watershed and aquifer drinking-water supplies and in built water and wastewater infrastructure in the Greater Chicago Area. The focus of chapter 1 is on assessing point-of-use (tapwater) drinking-water exposure pathways for a broad range of potential environmental contaminants and exploring infrastructure-related factors that could alter or transform chemical constituents or microbial communities in drinking water (such as treatment plant technology, distribution system characteristics, private plumbing components, and point-of-use treatment)

Pennsylvania has the second highest number of residential wells of any state in the Nation with approximately 2.4 million residents that depend on groundwater for their domestic water supply. The groundwater used for domestic water supply in Bradford County is obtained primarily from shallow bedrock and from unconsolidated (glacial) deposits that overlie the bedrock. As reported in chapter 2, data for 72 domestic wells were collected and analyzed for a wide range of constituents that could be evaluated in relation to drinking water health standards, geology, land use, and other environmental factors.

The occurrence of arsenic and uranium in groundwater at concentrations that exceed drinking-water standards is a concern because of the potential adverse effects on human health. The Connecticut Department of Public Health reported that there are about 322,600 private wells in Connecticut. The State does not require that existing private wells be routinely tested for arsenic, uranium, or other contaminants. The U.S. Geological Survey (USGS) completed an assessment in 2016 on the distribution of concentrations of arsenic and uranium in groundwater from bedrock in Connecticut. Chapter 3 presents the major findings for arsenic and uranium concentrations from water samples collected from 2013 to 2015 from private wells.

The circumstances and response to Flint's drinking water contamination involved implementation and oversight lapses at the EPA, the state of Michigan, the Michigan

Department of Environmental Quality (MDEQ), and the city of Flint. Chapter 4 evaluates additional matters concerning the agency's management controls when responding to the Flint contamination incident.

Federal agencies have identified several billion dollars in existing and future tribal drinking water and wastewater infrastructure needs. Chapter 5 examines the extent to which selected federal agencies identified tribes' drinking water and wastewater infrastructure needs and funded tribal water infrastructure projects, including tribes' most severe sanitation deficiencies.

Chapter 1 - The U.S. Geological Survey (USGS) Environmental Health Mission Area (EHMA) is providing comprehensive science on sources, movement, and transformation of contaminants and pathogens in watershed and aquifer drinking-water supplies and in built water and wastewater infrastructure (referred to as the USGS Water and Wastewater Infrastructure project) in the Greater Chicago Area and elsewhere in the United States, to fill data gaps identified by stakeholders and collaborators in drinking water and public health. EHMA Water and Wastewater Infrastructure research specifically provides insight into natural factors in the environment as well as those water-infrastructure components and processes (such as source-water corrosivity, treatment, plumbing, and so forth) that might influence human exposure to chemical and microbial contaminants at the residential tap. This infrastructure-exposure research role is fulfilled uniquely by the USGS and not by the U.S. Environmental Protection Agency (EPA), other agencies, or municipalities that focus on regulatory and policy activities and related compliance. The USGS approach to assessing the possible links between human health and chemical contaminant and pathogen exposure in drinking water is conducted in collaboration with public health experts and includes comprehensive characterization of the presence/absence and concentrations of more than 500 organic and 27 inorganic chemical constituents at the point of use (tap). Laboratory results for lead and other inorganic contaminants in Chicago, Illinois, and East Chicago, Indiana, residential tapwater are being released to ensure the timely release of quality-assured data to participants in the study. Concentrations of lead and other inorganic constituents were assessed in drinking water at the point of use (kitchen tap or filter) in 45 residential locations and in two locations within each of the two Chicago water purification plants and the two East Chicago water filtration plants during July–December 2017. Three methods were used for analyzing lead. The most sensitive method had a reporting limit of 0.020 micrograms per liter (μg/L). When using the most sensitive analytical method, lead was detected in 39 of 45 residential tapwater samples, with concentrations ranging from less than 0.020 μg/L to 5.31 μg/L (median of the detected values = 0.481 μg/L). Concentrations of lead also were detected in Lake Michigan intake water at all water purification/filtration plant facilities at concentrations ranging from 0.083 to 0.330 μg/L, but were not detected above the reporting limit in any samples of treated, pre-distribution drinking water at any of the water purification/filtration plant facilities. Because the USGS Water and Wastewater

Infrastructure project in the Greater Chicago Area is focused on the potential human exposure to a broad suite of organic and inorganic contaminants in drinking water and is not focused specifically on lead, the sampling protocol did not include "first-draw," stagnant sampling and samples were collected with point-of-use treatment in place, if present. *Thus, the lead results reported herein are not appropriate for assessment of compliance with the EPA 1991 Lead and Copper Rule.* Information resources for lead mitigation and water filtration are provided.

Chapter 2 - Pennsylvania has the second highest number of residential wells of any state in the Nation with approximately 2.4 million residents that depend on groundwater for their domestic water supply. Despite the widespread reliance on groundwater in rural areas of the state, publicly available data to characterize the quality of private well water are limited. In Bradford County, more than half of the residents use groundwater from private domestic-supply wells as their primary drinking source. The quality of private well water is influenced by the regional and local setting, including the surrounding soil, geology, land use, household plumbing, and well construction. The groundwater used for domestic water supply in Bradford County is obtained primarily from shallow bedrock and from unconsolidated (glacial) deposits that overlie the bedrock. Historical land use has been predominately forested, agricultural, and residential, but more recently unconventional oil/gas development has been distributed throughout the landscape. Pennsylvania is one of only two states in the Nation without statewide water-well construction standards. To better assess the quality of groundwater used for drinking water supply in Bradford County, data for 72 domestic wells were collected and analyzed for a wide range of constituents that could be evaluated in relation to drinking water health standards, geology, land use, and other environmental factors. Groundwater samples were collected from May through August 2016 and analyzed for physical and chemical properties, including major ions, nutrients, trace elements, volatile organic compounds, ethylene and propylene glycol, alcohols, gross-alpha/beta-particle activity, uranium, radon-222, and dissolved gases. A subset of samples was analyzed for radium isotopes (radium-226 and -228) and for the isotopic composition of methane. This study was conducted by the U.S. Geological Survey in cooperation with the Northern Tier Regional Planning and Development Commission and is part of a regional effort to characterize groundwater in rural areas of Pennsylvania. Results of the 2016 study show that groundwater quality generally met most drinking-water standards. However, a percentage of samples failed to meet maximum contaminant levels (MCLs) for total coliform bacteria (49.3 percent), *Escherichia coli* (8.5 percent), barium (2.8 percent), and arsenic (2.8 percent); and secondary maximum contaminant levels (SMCL) for sodium (48.6 percent), manganese (30.6 percent), gross alpha and beta activity (16.7 percent), iron (11.1 percent), pH (8.3 percent), total dissolved solids (5.6 percent), chloride (1.4 percent), and aluminum (1.4 percent). Radon-222 activities exceeded the proposed drinking-water standard of 300 picocuries per liter (pCi/L) in 70.4 percent of the samples.

There were no exceedances of drinking water health standards for any volatile organic compounds, and the only detections were for three trihalomethanes in one sample. The pH of the groundwater had a large influence on chemical characteristics and ranged from 6.18 to 9.31. Generally, the higher pH samples had higher potential for elevated concentrations of several constituents, including total dissolved solids, sodium, lithium, chloride, fluoride, boron, arsenic, and methane. For the Bradford County well-water samples, calcium/bicarbonate type waters were most abundant, with others classified as sodium/bicarbonate or mixed water types including calcium-sodium/bicarbonate, calcium-sodium/ bicarbonate-chloride, sodium/bicarbonate-chloride, sodium/ bicarbonate-sulfate, or sodium/chloride types. Six principal components (pH, redox, hardness, chloride-bromide, strontium-barium, and molybdenum-arsenic) explained nearly 78.3 percent of the variance in the groundwater dataset. Groundwater from 12.5 percent of the wells had concentrations of methane greater than the Pennsylvania action level of 7 milligrams per liter (mg/L); detectable methane concentrations ranged from 0.01 to 77 mg/L. In addition, low levels of ethane (as much as 0.13 mg/L) were present in seven samples with the highest methane concentrations. The isotopic composition of methane in five of these groundwater samples was consistent with the isotopic compositions reported for mud-gas logging samples from these geologic units and a thermogenic source. Isotopic composition from a sixth sample suggested the methane in that sample may be of microbial origin. Well-water samples with the higher methane concentrations also had higher pH values and elevated concentrations of sodium, lithium, boron, fluoride, arsenic, and bromide. Relatively elevated concentrations of some other constituents, such as barium and chloride, commonly were present in, but not limited to, those well-water samples with elevated methane. Four of the six groundwater samples with the highest methane concentrations had chloride/bromide ratios that indicate mixing with a small amount of brine (0.02 percent or less) similar in composition to those reported for gas and oil well brines in Pennsylvania. In several other eastern Pennsylvania counties where gas drilling is absent, groundwater with comparable chloride/bromide ratios and chloride concentrations have been reported, implying a potential natural source of brine. Most of Bradford County well-water samples have chloride concentrations less than 20 mg/L, and those with higher chloride concentrations have chloride/bromide ratios that indicate anthropogenic sources (such as road-deicing salt and septic effluent) or brine. Brines that are naturally present may originate from deeper parts of the aquifer system, whereas anthropogenic sources are more likely to affect shallow groundwater because they occur on or near the land surface. The available data for this study indicate that no one physical factor, such as the topographic setting, well depth, or altitude at the bottom of the well, was particularly useful for predicting those well locations with an elevated dissolved concentration of methane. The 2016 assessment of groundwater quality in Bradford County shows groundwater is generally of good quality, but methane and some constituents that occur in high concentration in naturally occurring brine and also in produced waters may be present at low to moderate concentrations in groundwater in various parts of the aquifer.

Chapter 3 - The occurrence of arsenic and uranium in groundwater at concentrations that exceed drinking-water standards is a concern because of the potential adverse effects on human health. Some early studies of arsenic occurrence in groundwater considered anthropogenic causes, but more recent studies have focused on sources of naturally occurring arsenic to groundwater, such as minerals within aquifer materials that are in contact with groundwater. Arsenic and uranium in groundwater in New England have been shown to have a strong association to the geologic setting and nearby streambed sediment concentrations. In New Hampshire and Massachusetts, arsenic and uranium concentrations greater than human-health benchmarks have shown distinct spatial patterns when related to the bedrock units mapped at the local scale. The Connecticut Department of Public Health reported that there are about 322,600 private wells in Connecticut serving approximately 823,000 people, or 23 percent of the State's population. The State does not require that existing private wells be routinely tested for arsenic, uranium, or other contaminants; consequently, private wells are only sampled at the well owner's discretion or when they are newly constructed. The U.S. Geological Survey (USGS), in cooperation with the DPH, completed an assessment in 2016 on the distribution of concentrations of arsenic and uranium in groundwater from bedrock in Connecticut. This report presents the major findings for arsenic and uranium concentrations from water samples collected from 2013 to 2015 from private wells.

Chapter 4 - The U.S. Environmental Protection Agency's (EPA's) Office of Inspector General (OIG) conducted this audit to examine the circumstances of, and the EPA's response to, contamination in the city of Flint, Michigan's community water system, including the EPA's exercise of its oversight authority. After Flint switched its drinking water supply in April 2014, inadequate water treatment exposed many Flint residents to lead. In October 2016, the OIG issued a Management Alert to the EPA about specific factors that delayed its emergency response. Our current report evaluates additional matters concerning the agency's management controls when responding to the Flint contamination incident. According to the MDEQ, more than $400 million in public funds has been spent by the EPA and Michigan to address the Flint crisis. Eight other federal agencies also provided support.

Chapter 5 - Tens of thousands of American Indians and Alaska Natives do not have safe drinking water or wastewater disposal in their home—referred to as needs arising from a sanitation deficiency—at a higher percentage than the general population, according to IHS. Among other things, IHS assesses homes, either individually or by reviewing public water systems, to determine any deficiencies. Seven agencies, including IHS, EPA, and USDA, have programs that provide drinking water and wastewater infrastructure assistance to Indian tribes. GAO was asked to review federal efforts to provide water infrastructure assistance to Indian tribes. This report examines, among other objectives, the extent to which selected federal agencies (1) identified tribes' drinking water and wastewater infrastructure needs and (2) funded tribal water infrastructure projects, including tribes' most severe sanitation deficiencies. GAO reviewed agency data on tribal needs, analyzed agency funding data for tribal water infrastructure projects, reviewed agency policy documents, and interviewed agency officials and officials from 22 tribes representing different geographic locations.

In: Drinking Water
Editor: Erwin Jon Orosz
ISBN: 978-1-53617-284-3

Chapter 1

Concentrations of Lead and Other Inorganic Constituents in Samples of Raw Intake and Treated Drinking Water From the Municipal Water Filtration Plant and Residential Tapwater in Chicago, Illinois, and East Chicago, Indiana, July–December 2017*

Kristin M. Romanok, Dana W. Kolpin, Shannon M. Meppelink, Michael J. Focazio, Maria Argos, Mary E. Hollingsworth, R. Blaine McCleskey, Andrea R. Putz, Alan Stark, Christopher P. Weis, Abderrahman Zehraoui and Paul M. Bradley

Abstract

The U.S. Geological Survey (USGS) Environmental Health Mission Area (EHMA) is providing comprehensive science on sources, movement, and transformation of contaminants and pathogens in watershed and aquifer drinking-water supplies and in built water and wastewater infrastructure (referred to as the USGS Water and Wastewater Infrastructure project) in the Greater Chicago Area and elsewhere in the United States, to fill data gaps identified by stakeholders and collaborators in drinking water and public health. EHMA Water and Wastewater Infrastructure research specifically provides insight into natural factors in the environment as well as those water-infrastructure components and processes (such as source-water corrosivity, treatment, plumbing, and so

* This is an edited, reformatted and augmented version of a U.S. Department of the Interior and U.S. Geological Survey, Open-File Report 2018–1071, dated July–December 2017.

forth) that might influence human exposure to chemical and microbial contaminants at the residential tap. This infrastructure-exposure research role is fulfilled uniquely by the USGS and not by the U.S. Environmental Protection Agency (EPA), other agencies, or municipalities that focus on regulatory and policy activities and related compliance. The USGS approach to assessing the possible links between human health and chemical contaminant and pathogen exposure in drinking water is conducted in collaboration with public health experts and includes comprehensive characterization of the presence/absence and concentrations of more than 500 organic and 27 inorganic chemical constituents at the point of use (tap).

Laboratory results for lead and other inorganic contaminants in Chicago, Illinois, and East Chicago, Indiana, residential tapwater are being released to ensure the timely release of quality-assured data to participants in the study. Concentrations of lead and other inorganic constituents were assessed in drinking water at the point of use (kitchen tap or filter) in 45 residential locations and in two locations within each of the two Chicago water purification plants and the two East Chicago water filtration plants during July–December 2017. Three methods were used for analyzing lead. The most sensitive method had a reporting limit of 0.020 micrograms per liter (µg/L). When using the most sensitive analytical method, lead was detected in 39 of 45 residential tapwater samples, with concentrations ranging from less than 0.020 µg/L to 5.31 µg/L (median of the detected values = 0.481 µg/L). Concentrations of lead also were detected in Lake Michigan intake water at all water purification/filtration plant facilities at concentrations ranging from 0.083 to 0.330 µg/L, but were not detected above the reporting limit in any samples of treated, pre-distribution drinking water at any of the water purification/filtration plant facilities.

Because the USGS Water and Wastewater Infrastructure project in the Greater Chicago Area is focused on the potential human exposure to a broad suite of organic and inorganic contaminants in drinking water and is not focused specifically on lead, the sampling protocol did not include "first-draw," stagnant sampling and samples were collected with point-of-use treatment in place, if present. *Thus, the lead results reported herein are not appropriate for assessment of compliance with the EPA 1991 Lead and Copper Rule.* Information resources for lead mitigation and water filtration are provided.

Introduction

Population growth, increased urbanization, complex environmental contaminant mixtures, and aging water-supply infrastructure have raised public concerns about the quality and long-term sustainability of drinking water across the United States. For decades, the U.S. Geological Survey (USGS) has assessed the quantity and quality of the Nation's ambient groundwater and surface-water drinking-water supplies (https://www.usgs.gov/science/mission-areas). To help address the public's concerns about water quality and to provide regulatory and public health agencies the scientific context for managing human-health risks associated with exposure to contaminants in drinking water, the USGS Environmental Health Mission Area (EHMA, https://www2.usgs.gov/envirohealth/) has initiated water and wastewater infrastructure research. The focus of the research is on (1) assessing point-of-use (tapwater) drinking-water exposure pathways for a broad range of potential environmental contaminants and

(2) exploring infrastructure-related factors that could alter or transform chemical constituents or microbial communities in drinking water (such as treatment plant technology, distribution system characteristics, private plumbing components, and point-of-use treatment) (Figure 1).

The USGS EHMA is collaborating with the University of Illinois at Chicago (UIC) School of Public Health and the National Institutes of Health/National Institute of Environmental Health Sciences (NIH/NIEHS) to assess the quality and potential human-health concerns of residential tapwater in the Greater Chicago area, including in the city of East Chicago, Indiana. The USGS EHMA is focusing on developing comprehensive datasets that document the presence and (or) absence and concentrations of more than 500 organic and 27 inorganic chemical constituents as well as conducting analyses to determine the presence of bacterial community composition in tapwater to provide insights into infrastructure-related factors that might influence contaminant exposures at the point of use. These datasets are scoped and collected in collaboration with UIC and NIH/NIEHS to explore associations between contaminant exposures in drinking water and human health. The overall project objective is to characterize the occurrence of a broad range of possible organic and inorganic constituents in tapwater and possible links to human health. Lead (Pb) contamination of tapwater is a concern for residents of Chicago, Illinois, and East Chicago, Indiana, because of the existence of legacy lead service lines in both public and private (curb to tap premise plumbing) segments of the municipal drinking-water distribution systems, the documented occurrence of low orthophosphate levels in the drinking-water system, the documented industrial lead contamination of surface soils and ongoing Superfund assessment and mitigation in East Chicago, and concerns that active remediation of soil contamination may alter the structural integrity and biochemical quality of the drinking-water supply in East Chicago (EPA, 2018d). Thus, the timely release of quality-assured data on concentrations of Pb and other inorganic constituents in tapwater samples collected as part of the EHMA's Water and Wastewater Infrastructure Project's Greater Chicago Area investigation is an important part of the EHMA's Tapwater Exposure Study.

Background

The U.S. Environmental Protection Agency (EPA) has established human-health regulations and guidelines for inorganic drinking-water constituents under the National Primary Drinking Water Regulations (NPDWR) (EPA, 2018c) and the Lead and Copper Rule (LCR) of 1991 (EPA, 2008, 2018a). These standards include Maximum Contaminant Level (MCL), Maximum Contaminant Level Goal (MCLG), and Treatment Technique Action Level (TTAL) concentration standards to limit contaminants in public drinking water. Compliance monitoring in regard to these standards is based on established EPA sample-collection methods (EPA, 2016). In the current report, reference to these standards is for comparison purposes only.

The TTAL for Pb is 15 micrograms per liter (μg/L) (additional information available at EPA, 2008, 2018a). Exceedance of the TTAL for samples collected using EPA protocols may require additional tapwater quality monitoring, corrosion control treatment, source water monitoring/ treatment, public education, and replacement of lead service lines (EPA, 2008). For example, the City of East Chicago currently (2018) uses orthophosphate at an average concentration of 1.35 milligrams per liter (mg/L) for corrosion inhibition (Indiana General Assembly, 2018) and, under a fund from the Indiana Department of Environmental Management (IDEM), is replacing lead service lines in three Superfund zones to address concerns about Pb in tapwater. The NPDWR MCLG is defined as the level in drinking water below which there is no known or expected risk to health and is a non- enforceable public health goal (EPA, 2018c). The EPA has set the MCLG for Pb at zero (EPA, 2018c), "because lead is a toxic metal that can be harmful to human health even at low exposure levels" (EPA, 2018b).

The LCR is intended to minimize Pb levels in drinking water primarily through corrosion control and monitoring of community-water-system tapwater samples (EPA, 2008, 2018a). The monitoring of Pb contaminants in tapwater under the LCR is focused on the protection of human health, including vulnerable populations (infants, young children, pregnant women), from possible contamination related to Pb-based materials (Pb pipes, Pb solder, and so forth) in the drinking-water distribution system (EPA, 2008, 2017, 2018a, 2018c; Triantafyllidou and Edwards, 2012). Lead contaminants originating from Pb-based plumbing materials can leach into drinking water, leading to increasing dissolved concentrations under static conditions over time (Triantafyllidou and Edwards, 2012). Consequently, the protocol for LCR compliance monitoring is intended to target a presumptive "worst-case" exposure (EPA, 2008, 2016; Triantafyllidou and Edwards, 2012). Guidance for LCR compliance monitoring stipulates collection of "first-draw," stagnant (minimum of 6 hours since the last draw, typically overnight) samples of cold tapwater from a faucet without point-of-use treatment in homes in the service area that are considered to be at greatest risk of elevated Pb concentrations (EPA, 2016).

In contrast, the USGS EHMA Greater Chicago Area Tapwater Exposure Study is being conducted to assess the potential human exposure to a broad suite of organic and inorganic contaminants in drinking water. This study is not focused specifically on Pb, and the sampling protocol does not include first-draw, stagnant sampling. All samples were collected with faucet screens (aerator screens) and point-of-use treatment equipment left in place. Domestic tapwater sample collections were distributed throughout the day, resulting in a range of pre-sample flushing periods. Information on resident water use prior to the beginning of sampling was not obtained. Trace element samples were the 11th (USGS National Water Quality Laboratory, NWQL) and 12th (Redox Chemistry Laboratory, RCL) containers in the sample collection sequence and were collected approximately 5 minutes after the faucet was turned on. Considering only the water used during the collection process, the USGS protocol involves flushing approximately 12 liters (3 gallons) of water prior to trace element sample collection. The tapwater Pb results reported herein are not appropriate for assessment of LCR compliance, because the samples were not collected using a first-draw, stagnant sampling protocol and, where present, were collected with the point-of-use treatment in place.

Purpose and Scope

The purpose of this report is to describe the site selection, sampling protocol, and inorganic chemical analytical methods used for the collection of samples of residential tapwater as well as samples of untreated (raw intake water) and treated (finished water) pre-distribution water from two Chicago water purification plants and two East Chicago water filtration plants during the EHMA's Water and Wastewater Infrastructure Project's Greater Chicago Area Tapwater Exposure Study. Samples discussed in this report were collected in Chicago, Illinois, and East Chicago, Indiana, during July–December 2017. This sampling was planned and implemented in cooperation with the UIC School of Public Health, City of Chicago Department of Water Management, City of East Chicago Utilities Department, and other non-governmental organizations. EPA regulations presented in this report are provided for reference only.

METHODS

Site selection, sampling protocols, and analytical methods are described in this section.

Sampling Site Selection

Sampling sites were selected on the basis of the availability and permission of Chicago and East Chicago community volunteers. Sites in Chicago were selected to ensure equal number of samples and broad spatial coverage of residences within the service areas of the two water filtration plants. Sites in East Chicago were selected to ensure samples were collected from residences within the area of documented soil Pb contamination and to provide broad spatial coverage of the East Chicago study area. One-time site access of approximately 30 minutes was required to complete the sampling of the domestic water supply at a kitchen tap. Samples were collected during July–December 2017 from a total of 45 residences (Table 1). All domestic tapwater station names, station numbers, location information, and collection times and dates are anonymized. Latitude and longitude data are generalized and do not indicate exact sampling site location. Additionally, for comparison purposes, water-quality samples were collected from the raw-water intake in Lake Michigan and from post-treatment (pre-distribution) locations two times each at the Chicago (July, November) water purification plant facilities and at the two East Chicago (August, September) water filtration plant facilities in operation at the time of residential sample collection (Table 1). A water-parcel-tracking (Lagrangian) sampling approach was not used in this study. In other words, collections were not timed to sample a theoretical parcel of water as it moved from the raw-water intake through the filtration plant water and distribution system to individual residences.

Sampling Methods

Domestic tapwater sampling site visits were distributed throughout the day, as dictated by residents' in-home availability. Detailed protocols, following guidelines established by the USGS (U.S. Geological Survey, variously dated), were provided to USGS personnel who conducted the sampling (Romanok and others, 2018b). An initial set of disposable gloves was donned to lay out a clean polyethylene (chamber bag) workspace on the floor in front of the kitchen sink and to record date and time on pre-labeled sample bottles. Fresh gloves were then donned to fill and cap narrow-neck sample bottles; sample collection began immediately after the faucet was turned on. Samples were collected with the faucet screen (aerator screen) in place. If the tap was equipped with a point-of-use treatment or other filtration device, the tapwater sample was collected post-treatment/filtration and the treatment device was visually inspected and noted along with the resident's recollection of the device maintenance history (Table 1). Approximately 30 containers for each sample were filled with cold tapwater in the same sequence established for previous USGS tapwater studies in order to improve comparison of results. After all samples were collected at a given site, field properties (dissolved oxygen [mg/L], water temperature [degrees Celsius], specific conductance [micro-siemens per centimeter at 25 degrees Celsius, μS/cm], and pH [standard units]) were measured in an open chamber with water flowing (Romanok and others, 2018b). Total sample collection time (including measurements of field properties) was approximately 20 minutes. No information was obtained about the composition or materials in residential (curb-to-tap) drinking-water service lines and plumbing.

Two samples were collected in polyethylene containers for trace element analysis (Fishman and Friedman, 1989; Hoffman and others, 1996; Hergenreder, 2011; USEPA, 2014). For the first sample, an acid-rinsed, 250-milliliter (mL), narrow-neck (28-mL) polyethylene bottle was pre-rinsed three times with sample water, and samples were collected as follows. Approximately 20 mL of tapwater was collected, capped, vigorously shaken, and discarded. The bottle was then filled to the shoulder with tapwater, preserved with 2 mL of 7.5 Normality, ultrapure grade, nitric acid (HNO), capped, shaken to completely mix, sealed in two self-closing plastic bags, and placed on ice. The trace element samples and other chemical analysis samples for each location were shipped together overnight, on ice, to the USGS Organic Geochemistry Research Laboratory (OGRL) in Lawrence, Kansas. All trace element samples were then aggregated into a single cooler and shipped overnight on ice to the USGS NWQL in Denver, Colorado. The second trace element sample was collected in a 15-mL polyethylene Falcon tube, acidified with 150 microliters of 7.5 Normality, ultrapure grade, nitric acid, similarly bagged, and shipped overnight to the RCL in Boulder, Colo. Because the trace element samples were the 11th and 12th containers in the sample collection sequence, these samples were collected approximately 5 minutes after the faucet was turned on and after approximately 12 liters (3 gallons) of water flow.

Analytical Methods

Samples from 45 residential kitchen taps and the four Chicago and four East Chicago water filtration plant locations (two locations in each of the four facilities) were analyzed at the NWQL and RCL for concentrations of trace elements by using three analytical methods. Results for all trace element analytes, in milligrams per liter and micrograms per liter, are available for download at (Romanok and others, 2018a). Lead concentrations are summarized in Table 1.

National Water Quality Laboratory

Whole-water samples were analyzed for 27 trace elements, including Pb, as described in (Fishman and Friedman, 1989; Hoffman and others, 1996). Briefly, samples were digested with hydrochloric acid, heated, and analyzed by inductively coupled plasma-mass spectrometry (ICP-MS) or by atomic absorption (cadmium and calcium) (as described in Fishman and Friedman, 1989; Hoffman and others, 1996). Trace element concentrations, including Pb, are reported in microgram per liter concentrations. The reporting level (RL) type used by the NWQL varies by analyte. The RL type used to report Pb concentrations is DLBLK, meaning the detection level is determined using blank data. A description of how this RL is determined can be found in U.S. Geological Survey (2015). The detection level for Pb, for this analytical method, was 0.020 μg/L.

Table 1. Concentration results of lead analyzed by the U.S. Geological Survey, National Water Quality Laboratory, Denver, Colorado, and Redox Chemistry Laboratory, Boulder, Colorado, for the U.S. Geological Survey Environmental Health Mission Area, Water and Wastewater Infrastructure Project, Tapwater Study, Chicago, Illinois, and East Chicago, Indiana, July–December, 201

Station number	State	Station name	Medium code	Sampling trip	Lead (µg/L, ICP-MS[1])	Lead (µg/L, GFAAS[2])	Lead (mg/L, ICP-OES[3])	Point-of-use treatment[2]
414500087320009	Illinois	38N-15E-29 001 (raw water)	WS	Trip 1	0.098	NA	NA	—
414500087320010	Illinois	38N-15E-29 002 (pre-distribution water)	WT	Trip 1	< 0.020	NA	NA	—
415300087360009	Illinois	39N-14E-10 001 (raw water)	WS	Trip 1	0.083	< 0.6	< 0.008	—
415300087360010	Illinois	39N-14E-10 002 (pre-distribution water)	WT	Trip 1	< 0.020	NA	NA	—
91808900008	Indiana	IN5245012 LAKE EAST CHICAGO WATER WORKS (WFP 1 pre-distribution)	WT	Trip 2	< 0.020	< 0.600	< 0.008	—
413700087270001	Indiana	37N 9W 28 003	WT	Trip 2	2.49	2.61	< 0.008	Yes
413700087270002	Indiana	37N 9W 33 004	WT	Trip 2	0.020	< 0.600	< 0.008	Yes
413700087280001	Indiana	37N 9W 29 002	WT	Trip 2	0.090	< 0.600	< 0.008	No
413700087280002	Indiana	37N 9W 28 007	WT	Trip 2	1.15	< 0.600	< 0.008	Yes
413800087260001	Indiana	37N 9W 22 006	WT	Trip 2	1.92	1.16	< 0.008	No
413800087260002	Indiana	37N 9W 27 012	WT	Trip 2	0.740	< 0.600	< 0.008	No
413800087270001	Indiana	37N 9W 21 013	WT	Trip 2	0.280	< 0.600	< 0.008	No
413800087270002	Indiana	37N 9W 22 005	WT	Trip 2	0.060	< 0.600	< 0.008	No
413800087280001	Indiana	37N 9W 29 001	WT	Trip 2	0.530	< 0.600	< 0.008	No
413900087260001	Indiana	IN5245012 LAKE 00087 INTAKE 1 LAKE MICHIGAN (WFP 1 raw water)	WS	Trip 2	0.130	< 0.600	< 0.008	—

Station number	State	Station name	Medium code	Sampling trip	Lead (µg/L, ICP-MS1)	Lead (µg/L, GFAAS2)	Lead (mg/L, ICP-OES3)	Point-of-use treatment2
413900087260003	Indiana	IN5245012 LAKE 05021 INTAKE 2 LAKE MICHIGAN (WFP 2 raw water)	WS	Trip 2	0.210	< 0.600	< 0.008	—
413900087260004	Indiana	37N 9W 22 008 (WFP 2 pre-distribution)	WT	Trip 2	< 0.020	< 0.600	< 0.008	—
91808900008	Indiana	IN5245012 LAKE EAST CHICAGO WATER WORKS (WFP 1 pre-distribution)	WT	Trip 3	< 0.020	< 0.600	< 0.008	—
413900087260001	Indiana	IN5245012 LAKE 00087 INTAKE 1 LAKE MICHIGAN (WFP 1 raw water)	WS	Trip 3	0.250	< 0.600	< 0.008	—
413900087260003	Indiana	IN5245012 LAKE 05021 INTAKE 2 LAKE MICHIGAN (WFP 2 raw water)	WS	Trip 3	0.330	< 0.600	< 0.008	—
413900087260004	Indiana	37N 9W 22 008 (WFP 2 pre-distribution)	WT	Trip 3	< 0.020	< 0.600	< 0.008	—
413700087260001	Indiana	37N 9W 27 018	WT	Trip 3	1.12	0.921	< 0.008	No
413700087270003	Indiana	37N 9W 33 015	WT	Trip 3	0.480	< 0.600	< 0.008	Yes
413700087270004	Indiana	37N 9W 27 017	WT	Trip 3	0.020	< 0.600	< 0.008	Yes
413800087260003	Indiana	37N 9W 22 016	WT	Trip 3	< 0.020	< 0.600	< 0.008	Yes
413800087260004	Indiana	37N 9W 27 019	WT	Trip 3	1.58	1.38	< 0.008	No
413800087280002	Indiana	37N 9W 29 014	WT	Trip 3	1.36	1.00	< 0.008	No
414100087320001	Illinois	37N 15E 18 021	WT	Trip 4	0.489	< 0.600	< 0.008	No
414200087400001	Illinois	37N 14E 7 023	WT	Trip 4	0.03	< 0.600	< 0.008	Yes

Table 1. (Continued)

Station number	State	Station name	Medium code	Sampling trip	Lead (µg/L, ICP-MS1)	Lead (µg/L, GFAAS2)	Lead (mg/L, ICP-OES3)	Point-of-use treatment2
414200087400002	Illinois	37N 14E 7 024	WT	Trip 4	5.31	5.46	< 0.008	No
414300087320001	Illinois	37N 15E 6 020	WT	Trip 4	3.74	3.28	< 0.008	No
414300087330001	Illinois	37N 15E 7 022	WT	Trip 4	0.404	< 0.600	< 0.008	No
414500087320010	Illinois	38N-15E-29 002 (pre-distribution water)	WT	Trip 4	< 0.020	< 0.600	< 0.008	—
415300087360009	Illinois	39N-14E-10 001 (raw water)	WS	Trip 4	0.104	0.52	< 0.008	—
415300087360010	Illinois	39N-14E-10 002 (pre-distribution water)	WT	Trip 4	< 0.020	< 0.600	< 0.008	—
415400087420001	Illinois	40N 13E 36 028	WT	Trip 4	1.78	1.51	< 0.008	No
415400087420002	Illinois	40N 13E 36 029	WT	Trip 4	1.34	1.77	< 0.008	No
415500087410001	Illinois	40N 13E 36 026	WT	Trip 4	0.031	< 0.600	< 0.008	No
415500087410002	Illinois	40N 13E 25 027	WT	Trip 4	1.53	1.30	< 0.008	No
415500087420001	Illinois	40N 13E 25 030	WT	Trip 4	0.344	< 0.600	< 0.008	No
415600087380001	Illinois	40N 14E 28 025	WT	Trip 4	0.133	< 0.600	< 0.008	No
415700087410001	Illinois	40N 13E 13 033	WT	Trip 4	0.8	0.94	< 0.008	No
415700087440001	Illinois	40N 13E 15 031	WT	Trip 4	1.75	1.43	< 0.008	No
415800087390001	Illinois	40N 14E 17 034	WT	Trip 4	< 0.020	< 0.600	< 0.008	No
415800087420001	Illinois	40N 13E 11 032	WT	Trip 4	3.43	3.10	< 0.008	No
414400087440001	Illinois	38N 13E 34 045	WT	Trip 5	0.481	< 0.600	< 0.008	No
414600087390001	Illinois	38N 14E 20 044	WT	Trip 5	< 0.020	< 0.600	< 0.008	No
414700087340001	Illinois	38N 14E 12 047	WT	Trip 5	0.4	< 0.600	<0.008	No
414700087350001	Illinois	38N 14E 11 048	WT	Trip 5	2.06	1.41	< 0.008	No
414700087350002	Illinois	38N 14E 11 049	WT	Trip 5	0.616	< 0.600	< 0.008	No
414700087350003	Illinois	38N 14E 14 050	WT	Trip 5	0.486	< 0.600	< 0.008	No

Station number	State	Station name	Medium code	Sampling trip	Lead (µg/L, ICP-MS1)	Lead (µg/L, GFAAS2)	Lead (mg/L, ICP-OES3)	Point-of-use treatment2
414700087350004	Illinois	38N 14E 12 051	WT	Trip 5	< 0.020	< 0.600	< 0.008	No
414800087350001	Illinois	38N 14E 11 052	WT	Trip 5	0.0200	< 0.600	< 0.008	No
414800087360001	Illinois	38N 14E 3 043	WT	Trip 5	0.03	< 0.600	< 0.008	No
414800087360002	Illinois	38N 14E 11 046	WT	Trip 5	1.63	1.00	< 0.008	No
414900087360001	Illinois	38N 14E 2 042	WT	Trip 5	0.047	< 0.600	< 0.008	No
415400087410001	Illinois	39N 13E 01 039	WT	Trip 5	< 0.020	< 0.600	< 0.008	No
415500087460001	Illinois	40N 13E 32 038	WT	Trip 5	3.37	7.60	< 0.008	No
415800087390002	Illinois	40N 14E 08 040	WT	Trip 5	< 0.020	< 0.600	< 0.008	No
420000087380001	Illinois	41N 14E 32 041	WT	Trip 5	0.032	< 0.600	< 0.008	No

[1] Fishman and Friedman (1989); Hoffman and others (1996).

[2] Hergenreder (2011).

[3] EPA (2014).

[Samples were collected, during five trips, as part of the U.S. Geological Survey Environmental Health Mission Area, Water and Wastewater Infrastructure Project, Tapwater Study, East Chicago, Indiana, July–December 2017. Reporting levels: ICP-MS, 0.020 µg/L; GFAAS, 0.600 µg/L; ICP-OES, 0.008 mg/L. Abbreviations: ICP-MS, inductively coupled plasma-mass spectrometry; ICP-OES, inductively coupled plasma-optical emission spectrometry; GFAAS, graphite furnace atomic absorption spectrometry; mg/L, milligrams per liter; µg/L, micrograms per liter; WFP, water filtration plant; WT, treated water; WS, surface water; NA, not analyzed; <, less than; —, not relevant]

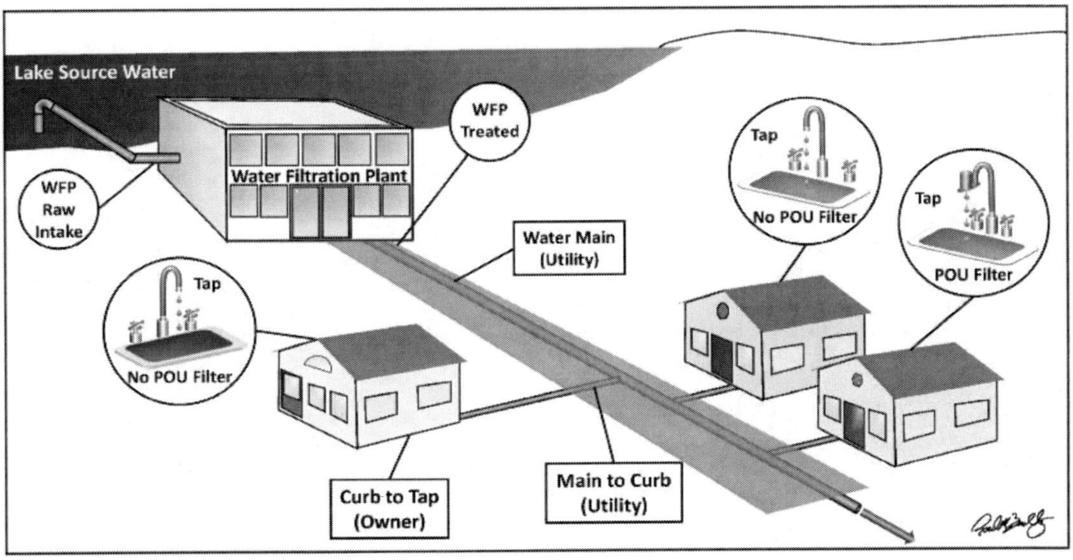

Figure 1. Typical water distribution system from untreated source water, through water filtration plant (WFP) to treated residential tap water, with and without point-of-use (POU) filtration device.

Redox Chemistry Laboratory

Whole-water samples were analyzed for 32 inorganic constituents and alkalinity at the RCL. Cations and trace elements were analyzed using inductively coupled plasma-optical emissions spectrometry (ICP-OES); anions were analyzed using ion chromatography; and alkalinity was determined using an automated titration, with first derivative near pH 4.5 (Pfaff, 1993; EPA, 2014). Trace element cations (including Pb) and anions are reported in milligrams per liter, and alkalinity is reported in milligrams per liter of bicarbonate. Concentrations of lead were also analyzed by graphite-furnace, atomic-absorption spectrometry (GFAAS), and results are reported in micrograms per liter. The method detection limit (MDL) (Childress and others, 1999; EPA, 2005) for Pb by GFAAS was 0.600 μg/L. The MDL for Pb, using the ICP-OES method, was 0.008 mg/L.

Quality Assurance/Quality Control

Field blanks were collected as part of the quality assurance/quality control (QAQC) protocols at three sites (station numbers: 414500087320010, 91808900008, 413900087260001) and analyzed at the NWQL and RCL to determine if any contamination bias was introduced during sampling. Field blank samples analyzed at the NWQL had detections for two trace elements, copper (0.400 μg/L) and silver (0.200 μg/L) (Romanok and others, 2018a). Reporting limits for copper and silver are 0.200 and 0.030 μg/L, respectively. Additionally, field blanks analyzed at RCL had detections for alkalinity (6.0 mg/L, as bicarbonate) and calcium (0.013 and 0.021 mg/L) (Romanok and others, 2018a). Method detection limits for alkalinity and calcium are 2.00 mg/L, as bicarbonate, and 0.006 mg/L, respectively. Lead was not detected in any field blank, using any method.

RESULTS

Lead concentration results are summarized in Table 1. Concentration data for all analyzed trace elements and alkalinity are provided in Romanok and others (2018a). Inorganic constituents regulated under the NPDWR or LCR are summarized in Table 2. Respective standards are listed for reference only.

Chicago and East Chicago Water Filtration Plant Sampling

Four water purification/filtration plant facilities (two in Chicago; two in East Chicago) supplied drinking water to residences included in 2017 Chicago Area Tapwater Exposure Study. Samples of raw water from Lake Michigan (7 total samples) and treated, pre-distribution drinking water (8 total samples) were collected at all four facilities during the residential tapwater sampling period. Lead, which has an RL of 0.020 µg/L, was detected only in raw-water intake samples (at all four facilities, in all collected samples) and only by using the most sensitive NWQL method (ICP-MS) (Fishman and Friedman, 1989; Hoffman and others, 1996) (Table 1). Lake Michigan raw-water intake Pb concentrations ranged from 0.083 to 0.330 µg/L, with a median of 0.130 µg/L (Table 1). Lead was not detected in any samples of treated, pre-distribution drinking water at either facility. No other inorganic constituents were detected in raw intake water or in treated, pre-distribution drinking water in excess of the respective EPA MCLG (public health goal) (Table 2).

Chicago Area Tapwater Sampling

Lead was detected at or above the RL (0.020 µg/L) in 39 (87 percent) of the 45 residential tapwater samples collected for this study (Table 1).

Table 2. Maximum concentrations of trace elements detected in tapwater for the U.S. Geological Survey Environmental Health Mission Area, Water and Wastewater Infrastructure Project, Tapwater Study, Chicago, Illinois, and East Chicago, Indiana, July–December, 2017

Trace element	MCL[1] (mg/L)	Public health goal[1] (mg/L)	Maximum value, RCL (mg/L)	Maximum value, NWQL[2] (mg/L)
Antimony	0.006	0.006	<0.030	0.00019
Arsenic	0.010	zero	< 0.040	0.001
Barium	2.00	2.00	0.108	0.022
Beryllium	0.004	0.004	<0.0005	0.00004
Cadmium	0.005	0.005	<0.003	0.00009
Chromium	0.100 (total)	0.100	0.002	0.0007
Copper	1.30 TTAL	1.30	0.019	0.036
Fluoride	4.00	4.00	0.723	—
Lead	0.015 TTAL	zero	0.008[3]/<0.008[4]	0.005
Selenium	0.050	0.050	<0.040	0.0003
Thallium	0.002	0.0005	—	0.00002
Uranium	0.030	zero	<0.010	0.0004

[1] EPA (2009).

[2] Fishman and Friedman (1989); Hoffman and others (1996).

[3] Hergenreder (2011).

[4] EPA (2014).

[U.S. Environmental Protection Agency (EPA) standards and public health goals are listed for reference only. Abbreviations: MCL, EPA Maximum Contaminant Level; TTAL, EPA Treatment Technique Action Level; —, not analyzed; mg/L, milligrams per liter; RLC, Redox Chemistry Laboratory; NWQL, National Water Quality Laboratory; <, less than]

Lead concentrations in tapwater ranged from below the reporting limit to 5.31 μg/L, with a median of 0.481 μg/L. The MCLG for Pb is zero, and 39 of 45 tapwater samples had detectable Pb concentrations when using the most sensitive method of analysis. No other inorganic constituents were detected in tapwater in excess of the corresponding EPA MCLG (public health goal) (Table 2).

DISCUSSION

The steps suggested by the EPA (2018a) to reduce Pb exposure in home drinking water include installation and proper maintenance of a point-of-use (at the tap) water filter that is certified to remove Pb. Six of the 15 East Chicago homes and one of the 30 Chicago homes sampled for this study had point-of-use (at the tap) water filters installed at the time of sampling, and some of the lowest tapwater Pb concentrations in the study were observed in four of these locations. The fifth highest concentration of Pb detected in the study (2.49 μg/L) corresponded to an unmaintained point-of- use filter unit.

The results of sampling reported for Pb and inorganic constituents in Chicago Area tapwater are only a subset of a broad suite of chemical constituents and microbial communities being assessed by the USGS as part of ongoing studies of tapwater in the Greater Chicago Area and elsewhere in the United States. The completed sample collections and datasets are intended to provide insight into those water infrastructure components and processes (such as source-water corrosivity, treatment, plumbing, and so forth) that might influence human exposure to chemical and microbial contaminants at the residential tap. Conducted in collaboration with colleagues and stakeholders from academia, public health agencies, and water utilities, this work can contribute to the foundational understanding of potential associations between human-health effects and drinking-water exposures at the tap.

EPA GUIDANCE ON REDUCING PB EXPOSURE IN HOME DRINKING WATER

The EPA lists five approaches to reduce exposure to Pb in home drinking water (EPA, 2018a).

- "Use only cold water for drinking, cooking and making baby formula. Boiling water does not remove lead from water.
- Regularly clean your faucet's screen (also known as an aerator).
- Consider using a water filter certified to remove lead and know when it's time to replace the filter.

- Before drinking, flush your pipes by running your tap, taking a shower, doing laundry or a load of dishes.
- Contact your water system to learn more about sources of lead and removing lead service lines."

CONVERSION FACTORS

U.S. customary units to International System of Units

Multiply	By	To obtain
Length		
inch (in.)	2.54	centimeter (cm)
inch (in.)	25.4	millimeter (mm)
foot (ft)	0.3048	meter (m)
mile (mi)	1.609	kilometer (km)
Area		
square mile (mi2)	2.590	square kilometer (km2)
Volume		
ounce, fluid (fl. oz)	0.02957	liter (L)

Temperature in degrees Celsius (°C) may be converted to degrees Fahrenheit (°F) as follows:

°F = (1.8 × °C) + 32.

Temperature in degrees Fahrenheit (°F) may be converted to degrees Celsius (°C) as follows:

°C = (°F − 32) / 1.8.

DATA

Horizontal coordinate information is referenced to the North American Datum of 1983 (NAD 83).

SUPPLEMENTAL INFORMATION

Specific conductance is given in microsiemens per centimeter at 25 degrees Celsius (μS/cm at 25°C).

Concentrations of chemical constituents in water are given either in milligrams per liter (mg/L) or micrograms per liter (μg/L).

REFERENCES

Childress, C. J. O., Foreman, W. T., Connor, B. F. & Maloney, T. J. (1999). New reporting procedures based on long-term method detection levels and some considerations for interpretations of water-quality data provided by the U.S. Geological Survey National Water Quality Laboratory: *U.S. Geological Survey Open-File Report*, 99–193, 19 p.

Fishman, M. J. & Friedman, L. C. eds. (1989). Methods for determination of inorganic substances in water and fluvial sediments: *U.S. Geological Survey Techniques of Water-Resources Investigations*, book 5, chap. A1, 545 p.

Hergenreder, R. L. (2011). *Trace metals in waters by GFAAS, in accordance with U.S. EPA and Health Canada requirements*: Waltham, Mass., PerkinElmer, Inc., 5 p.

Hoffman, G. L., Fishman, M. J. & Garbarino, J. R. (1996). Methods of analysis by the U.S. Geological Survey National Water Quality Laboratory; in-bottle acid digestion of whole-water samples: U.*S. Geological Survey Open-File Report*, 96–225, 28 p.

Indiana General Assembly. (2018). *Title 327. Water Pollution Control Division; Article 8. Public Water Supply; Rule 1. Public Water Supply Direct Additive and Indirect Additive Standards, Indiana Administrative Code 327–IAC–8*: Indiana General Assembly website, accessed March 4, 2018, at http://www.in.gov/legislative/iac/.

Pfaff, J. D. (1993). Method 300.0, *Determination of inorganic anions by ion chromatography, Revision 2.1: U.S. Environmental Protection Agency*, Environmental Monitoring Systems Laboratory, 30 p.

Romanok, K. M., Bradley, P. M., McCleskey, R. B., Meppelink, S. M., Argos, M. & Kolpin, D. W. (2018a). *Occurrence and concentrations of trace elements in discrete tapwater samples collected in Chicago, Illinois, and East Chicago, Indiana, 2017*: U.S. Geological Survey data release, https://doi.org/10.5066/F70R9NN0.

Romanok, K. M., Kolpin, D. W., Meppelink, S. M., Argos, M., Brown, J. B., DeVito, M. J., Dietze, J. E., Givens, C. E., Gray, J. L., Higgins, C. P., Hladik, M. L., Iwanowicz, L. R., Loftin, K. A., McCleskey, R. B., McDonough, C. A., Meyer, M. T., Strynar, M. J., Weis, C. P., Wilson, V. S. & Bradley, P. M. (2018b). Methods used for the collection and analyses of chemical and biological data for the Tapwater Exposure Study, United States, 2016–17: *U.S. Geological Survey Open-File Report 2018–1098*, 81 p., accessed November 1, 2018, at https:// doi.org/10.3133/ofr20181098.

Triantafyllidou, S. & Edwards, M. (2012). Lead (Pb) in tap water and in blood—Implications for lead exposure in the United States: *Critical Reviews in*

Environmental Science and Technology, v. 42, no. 13, p. 1297–1352, accessed March 4, 2018, at https://doi.org/10.1080/ 10643389.2011.556556.

U.S. Environmental Protection Agency [EPA]. (2008). Lead and Copper Rule—A quick reference guide: *EPA* 816–F–08–018, accessed March 4, 2018, at https://nepis.epa.gov/Exe/ ZyPDF.cgi?Dockey=60001N8P.txt.

U.S. Environmental Protection Agency [EPA]. (2011). Guidelines establishing test procedures for the analysis of pollutants: *Code of Federal Regulations*, Title 40, chap. 1, app. B to part 136—Definition and procedure for the determination of the method detection limit—Revision 1.11, p. 343–346, accessed March 4, 2018, at https://www.gpo.gov/fdsys/search/pagedetails.action?collectionCode=CFR& browsePath=Title+40%2FChapter+I%2FSubchapter+D%2FPart+136%2FAppendix+ B+to+Part+136&granuleId=CFR-2011-title40-vol23-part136- appB&packageId=CFR-2011-title40-vol23&collapse=true&fromBrowse=true.

U.S. Environmental Protection Agency [EPA]. (2014). *Inductively coupled plasma-optical emission spectrometry*, Method 6010D: EPA SW–846 Update V, Revision 4, 35 p., accessed November 2, 2017, at https:// www.epa.gov/sites/ production/files/2015-12/documents/6010d.pdf.

U.S. Environmental Protection Agency [EPA]. (2016). Quick guide to drinking water sample collection, 2d ed.: *U.S. Environmental Protection*, accessed February 26, 2018, at https://www.epa.gov/sites/ production/files/2015-11/documents/ drinking_water_sample_collection.pdf.

U.S. Environmental Protection Agency [EPA]. (2017). *Water quality standards: U.S. Code of Federal Regulations*, title 40, part 131, accessed March 4, 2018, at https://www.ecfr.gov/cgi-bin/text-idx?SID=454a7b51118b27f20cef29ff0 71c1440&node=40:22.0.1.1.18&rgn=div5.

U.S. Environmental Protection Agency [EPA]. (2018a). *USEPA Drinking Water Requirements for States and Public Water Systems*: Lead and Copper Rule website, accessed January 24, 2018, at https:// www.epa.gov/dwreginfo/lead-and-copper-rule#additional-resources.

U.S. Environmental Protection Agency [EPA]. (2018b). *Ground Water and Drinking Water: Basic Information about Lead in Drinking Water*: U.S. Environmental Protection Agency website, accessed January 24, 2018, at https://www.epa.gov/ ground-water-and-drinking-water/basic-information-about- lead-drinking-water#reducehome.

U.S. Environmental Protection Agency [EPA]. (2018c). *National Primary Drinking Water Regulations*: U.S. Environmental Protection Agency website, accessed January 11, 2018, at https://www.epa.gov/ground-water-and-drinking-water/national-primary-drinking-water-regulations.

U.S. Environmental Protection Agency [EPA]. (2018d). *USS lead drinking water pilot study*: U.S. Environmental Protection Agency website, accessed February 24, 2018, at https://www.epa.gov/uss-lead-superfund-site/uss-lead- drinking-water-pilot-study-data.

U.S. Geological Survey, variously dated, National field manual for the collection of water-quality data: U.*S. Geological Survey Techniques of Water-Resources Investigations*, book 9, chaps. A1–A9, accessed March 4, 2108, at https://pubs.water.usgs.gov/twri9A.

U.S. Geological Survey. (2015). Changes to National Water Quality Laboratory (NWQL) procedures used to establish and verify laboratory detection and reporting limits: *U.S. Geological Survey National Water Quality Laboratory Technical Memorandum* 15.02, accessed January 25, 2018, at https://nwql.usgs.gov/Public/tech_memos/ nwql.2015-02.pdf.

In: Drinking Water
Editor: Erwin Jon Orosz

ISBN: 978-1-53617-284-3

Chapter 2

DRINKING WATER HEALTH STANDARDS COMPARISON AND CHEMICAL ANALYSIS OF GROUNDWATER FOR 72 DOMESTIC WELLS IN BRADFORD COUNTY, PENNSYLVANIA, 2016*

John W. Clune and Charles A. Cravotta III

ABSTRACT

Pennsylvania has the second highest number of residential wells of any state in the Nation with approximately 2.4 million residents that depend on groundwater for their domestic water supply. Despite the widespread reliance on groundwater in rural areas of the state, publicly available data to characterize the quality of private well water are limited. In Bradford County, more than half of the residents use groundwater from private domestic-supply wells as their primary drinking source. The quality of private well water is influenced by the regional and local setting, including the surrounding soil, geology, land use, household plumbing, and well construction. The groundwater used for domestic water supply in Bradford County is obtained primarily from shallow bedrock and from unconsolidated (glacial) deposits that overlie the bedrock. Historical land use has been predominately forested, agricultural, and residential, but more recently unconventional oil/gas development has been distributed throughout the landscape. Pennsylvania is one of only two states in the Nation without statewide water-well construction standards.

To better assess the quality of groundwater used for drinking water supply in Bradford County, data for 72 domestic wells were collected and analyzed for a wide range of constituents that could be evaluated in relation to drinking water health standards, geology, land use, and other environmental factors. Groundwater samples were collected from May through August 2016 and analyzed for physical and chemical

* This is an edited, reformatted and augmented version of Open-File Report of U.S. Department of the Interior, U.S. Geological Survey, prepared in cooperation with the Northern Tier Regional Planning and Development Commission, Report 2018–5170 Version 1.2, dated May 2019.

properties, including major ions, nutrients, trace elements, volatile organic compounds, ethylene and propylene glycol, alcohols, gross-alpha/beta-particle activity, uranium, radon-222, and dissolved gases. A subset of samples was analyzed for radium isotopes (radium-226 and -228) and for the isotopic composition of methane. This study was conducted by the U.S. Geological Survey in cooperation with the Northern Tier Regional Planning and Development Commission and is part of a regional effort to characterize groundwater in rural areas of Pennsylvania.

Results of the 2016 study show that groundwater quality generally met most drinking-water standards. However, a percentage of samples failed to meet maximum contaminant levels (MCLs) for total coliform bacteria (49.3 percent), *Escherichia coli* (8.5 percent), barium (2.8 percent), and arsenic (2.8 percent); and secondary maximum contaminant levels (SMCL) for sodium (48.6 percent), manganese (30.6 percent), gross alpha and beta activity (16.7 percent), iron (11.1 percent), pH (8.3 percent), total dissolved solids (5.6 percent), chloride (1.4 percent), and aluminum (1.4 percent). Radon-222 activities exceeded the proposed drinking-water standard of 300 picocuries per liter (pCi/L) in 70.4 percent of the samples. There were no exceedances of drinking water health standards for any volatile organic compounds, and the only detections were for three trihalomethanes in one sample.

The pH of the groundwater had a large influence on chemical characteristics and ranged from 6.18 to 9.31. Generally, the higher pH samples had higher potential for elevated concentrations of several constituents, including total dissolved solids, sodium, lithium, chloride, fluoride, boron, arsenic, and methane. For the Bradford County well-water samples, calcium/bicarbonate type waters were most abundant, with others classified as sodium/bicarbonate or mixed water types including calcium-sodium/bicarbonate, calcium-sodium/ bicarbonate-chloride, sodium/bicarbonate-chloride, sodium/ bicarbonate-sulfate, or sodium/chloride types. Six principal components (pH, redox, hardness, chloride-bromide, strontium-barium, and molybdenum-arsenic) explained nearly 78.3 percent of the variance in the groundwater dataset.

Groundwater from 12.5 percent of the wells had concentrations of methane greater than the Pennsylvania action level of 7 milligrams per liter (mg/L); detectable methane concentrations ranged from 0.01 to 77 mg/L. In addition, low levels of ethane (as much as 0.13 mg/L) were present in seven samples with the highest methane concentrations. The isotopic composition of methane in five of these groundwater samples was consistent with the isotopic compositions reported for mud-gas logging samples from these geologic units and a thermogenic source. Isotopic composition from a sixth sample suggested the methane in that sample may be of microbial origin. Well-water samples with the higher methane concentrations also had higher pH values and elevated concentrations of sodium, lithium, boron, fluoride, arsenic, and bromide.

Relatively elevated concentrations of some other constituents, such as barium and chloride, commonly were present in, but not limited to, those well-water samples with elevated methane.

Four of the six groundwater samples with the highest methane concentrations had chloride/bromide ratios that indicate mixing with a small amount of brine (0.02 percent or less) similar in composition to those reported for gas and oil well brines in Pennsylvania. In several other eastern Pennsylvania counties where gas drilling is absent, groundwater with comparable chloride/bromide ratios and chloride concentrations have been reported, implying a potential natural source of brine. Most of Bradford County well-water samples have chloride concentrations less than 20 mg/L, and those with higher chloride concentrations have chloride/bromide ratios that indicate anthropogenic sources (such as road-deicing salt and septic effluent) or brine. Brines that are naturally present may originate from deeper parts of the aquifer system, whereas anthropogenic sources are more likely to affect shallow groundwater because they occur on or near the land surface.

The available data for this study indicate that no one physical factor, such as the topographic setting, well depth, or altitude at the bottom of the well, was particularly useful for predicting those well locations with an elevated dissolved concentration of methane. The 2016 assessment of groundwater quality in Bradford County shows groundwater is generally of good quality, but methane and some constituents that occur in high concentration in naturally occurring brine and also in produced waters may be present at low to moderate concentrations in groundwater in various parts of the aquifer.

INTRODUCTION

Pennsylvania has the second highest number of residential wells of any state in the Nation with approximately 2.4 million residents that depend on groundwater for their domestic water supply. Despite the widespread reliance on groundwater in rural areas of the state, publicly available data to characterize the quality of private well water are limited (Low and Chichester, 2006; Giddings, 2014; Johnson and Belitz, 2017). In Bradford County, more than half of the residents use groundwater from private domestic-supply wells as their primary drinking source (Pennsylvania Bureau of Topographic and Geologic Survey, written commun., 2017).

Groundwater from more than one in five private domestic-supply wells in the Nation potentially exceeds a human-health standard and yet, the potential health risks associated with domestic groundwater supplies are often understudied in comparison to the research and testing of public water supplies (DeSimone, 2009). Unlike public water-supply systems, managing domestic-supply wells in Pennsylvania is solely the responsibility of the homeowner. Acute gastrointestinal illnesses have been attributed to parasitic, viral, and bacterial pathogens in rural private well water (Raina and others, 1999; Murphy and others, 2016). Metals such as lead can be present in the aquifer or leached from plumbing components by corrosive groundwater, causing delayed mental and physical development (Brown and Margolis, 2012; Belitz and others, 2016), and long-term exposure to arsenic has been linked to skin and internal cancers (National Research Council, 1999). The presence of metals like iron and manganese in drinking water can be a nuisance owing to the unpleasant taste, odor, and color and often require additional maintenance and treatment of household plumbing systems (Penn State Extension, 2017). The complex mixture and breakdown of radioactive elements and volatile organic compounds (VOCs) in drinking water also has been shown to be carcinogenic (Otton, 1992; National Research Council, 1998; Hopke and others, 2000). Elevated levels of dissolved gases like methane can pose an explosive safety risk if private wells are not properly vented (Swistock and Sharpe, 2006).

The quality of private well water is influenced by the regional and local setting, including the surrounding soil, geology, land use, and household plumbing. In addition, improper well construction (such as, no grouting and [or] sanitary well cap) can facilitate the introduction of potential contaminants from the surface into private groundwater wells and local aquifers (Zimmerman and others, 2001; Simpson, 2004; Swistock and Sharpe, 2005; Giddings, 2014). Despite legislative attempts, Pennsylvania is one of two states without statewide water-well construction standards (Swistock, 2012; Wagner, 2012; LaRegina, 2013; Giddings, 2014; Pennsylvania General Assembly, 2015a, b, 2016).

Previous national and statewide studies that include Bradford County provide general characterizations of water quality for a bedrock and glaciated groundwater system. In glaciated settings of the United States, the quality of groundwater is vulnerable to dissolved solids from the natural geology, anthropogenic contaminants such as rising chloride levels, and nuisance constituents like iron (Warner and Ayotte, 2015). Sampling of private wells in Pennsylvania show bedrock geology and well construction have a significant influence on water quality, and common contaminants often include bacteria, low pH, arsenic, lead, nitrate, and radon (Swistock and others, 1993; Swistock and others, 2009).

Early regional groundwater assessments in the Bradford County area show comparatively moderate dissolved solids and hardness, with the most troublesome constituents being naturally occurring iron and manganese (Taylor and others, 1983). The predominant hydrogeochemical groundwater zones exhibit a sodium-chloride water type in the bedrock (restricted flow) and calcium/bicarbonate in the glaciated valleys (unrestricted flow) (Williams and others, 1998). Wells that have open holes that have been drilled in the restricted groundwater flow of bedrock formations can produce naturally elevated levels of chloride, barium, strontium, and radium (Williams and others, 1998; Yager and others, 2017).

More recent research that has furthered the understanding of local groundwater quality in Bradford County and surrounding counties has been a byproduct of the unconventional natural gas development in areas of Pennsylvania underlain by the Marcellus Shale since 2005 (U.S. Environmental Protection Agency, 2012b; Vidic and others, 2013; Soeder, 2017). Groundwater sampling from a subset of private domestic-supply wells before and after gas well drilling did not suggest major influences from gas well drilling, but did show that 40 percent of the samples failed at least one common drinking water health standard such as coliform bacteria (Boyer and others, 2012). Most well water chemistry in the region has been shown to be comparable to historical groundwater (American Petroleum Institute/America's Natural Gas Alliance, 2013) with some evidence of animal waste, septic effluent, or road salt but no detectable impact of flowback waters from gas well development (Reilly and others, 2015). Pre- and post-drilling comparisons show the quality of water from private domestic-supply wells is largely influenced by groundwater contact time (residence time) with the natural geology (Weston Solutions, 2012).

Freshwater is often found where rainwater has more recently flushed the shallow aquifer, whereas brackish water may be encountered in deeper formations that have the remains of connate (water trapped in the pores of sedimentary sediments when deposited) seawater (Lohman, 1939). Analysis of pre-drilling groundwater shows the natural migration of Appalachian Basin brines along natural pathways—such as faults, fractures, lineaments, and so forth—in the bedrock of some shallow aquifers (Warner and others, 2012; Llewellyn, 2014). Analysis of robust propriety groundwater datasets from the natural gas industry show common exceedances of drinking water health standards for sodium, manganese, iron, lithium, turbidity, low pH, arsenic, lead, and barium that are often related to the geological formation, water type, and topographic position (Siegel and others, 2015).

Naturally occurring thermogenic- and biogenic-derived dissolved hydrocarbon gases such as methane can also be common in groundwater from private domestic-supply wells (Molofsky and others, 2011, 2013; Heisig and Scott, 2013; Botner and others, 2018), although the potential for stray gas from natural gas extraction development is a concern (Osborn and others, 2011; Jackson and others, 2013; Heilweil and others, 2015; Llewellyn and others, 2015) especially in valleys and near faults (Wen and others, 2018).

The U.S. Geological Survey (USGS) has conducted a series of county groundwater studies since 2007 in order to provide publicly available data to better characterize the quality of groundwater in rural areas of Pennsylvania. Studies in Pike County and Sullivan County indicated few exceedances of the drinking water health standards, and samples that had elevated pH and methane often showed higher concentrations of sodium, lithium, boron, fluoride, and bromide (Senior, 2009, 2014; Sloto, 2013; Senior and Cravotta, 2017). Studies in Wayne County and Lycoming County presented similar results but also included bacteria sampling that indicated exceedances of the drinking water health standards and geochemical modeling that offered further insight into the effect that water-rock reactions, topographic position, and pH have on the regional groundwater chemistry (Senior and others, 2017; Gross and Cravotta, 2017).

The research described in this report is a continuation of these ongoing USGS county groundwater quality assessments in Pennsylvania characterizing the quality of groundwater in Bradford County and aims to further the national, statewide, and regional understanding of previous groundwater investigations. At the local level, the results will serve the needs of resource managers with a comparison of the quality of private well water with drinking water health standards and provide a conceptual model of the natural and anthropogenic influences on local groundwater quality. This study was conducted by the U.S. Geological Survey in cooperation with the Northern Tier Regional Planning and Development Commission and is part of a regional effort to characterize groundwater in rural areas of Pennsylvania.

Purpose and Scope

The purpose of this report is to (1) compare the groundwater quality from private domestic-supply wells to drinking-water standards, and (2) to provide a chemical analysis of the natural and anthropogenic influences on groundwater quality in Bradford County, Pennsylvania. These objectives were met by the collection of analytical data for groundwater from 72 domestic wells sampled throughout Bradford County, Pennsylvania, from May through August of 2016. Groundwater samples were analyzed for physical and chemical properties, including major ions, nutrients, trace elements, VOCs, ethylene and propylene glycol, alcohols, gross-alpha/ beta-particle activity, uranium, radon-222, and dissolved gases. A subset of samples was analyzed for radium isotopes (radium-226 and -228) and for the isotopic composition of methane. The measured concentrations of analyzed constituents in groundwater were compared to U.S. Environmental Protection Agency (EPA) drinking-water standards. The groundwater-quality data, summary statistics, and geochemical computations are presented to further the understanding of the current (2016) chemical quality of groundwater from freshwater aquifers used by private domestic-supply wells in Bradford County. The relations among observed groundwater-quality characteristics, geology, topographic setting, land use, and other environmental variables associated with sampled wells are evaluated to explain the variability in the quality of the groundwater.

Description of Study Area

Bradford County occupies approximately 1,147 square miles in the glaciated portion of the Appalachian Plateaus Physiographic Province in north-central Pennsylvania (Figure 1*A*). The geology of Bradford County is characterized as gently folded clastic sedimentary bedrock that is overlain in most places by unconsolidated glacial deposits and alluvium (less than 20 feet thick). The bedrock is composed primarily of shale, siltstone, and sandstone strata of Devonian (400 million years ago) to Pennsylvanian (300 million years ago) age, including (from youngest to oldest) the Allegheny and Pottsville Formation, Burgoon Sandstone, Huntley Mountain Formation, Catskill Formation, and Lock Haven Formation (Miles and others, 2001). Fractures in the bedrock provide water-bearing zones that store and transfer groundwater to wells. In the valleys, layers of fluvial outwash and lacustrine (lake) deposits from past glaciations have filled the lowlands with mostly unconsolidated silt, clay, sand, and gravel (Figure 1*C*). Figure 1*C* does not show lithology of surficial material or bedrock for the entire county which is often a mix of both. Private domestic-supply wells may be completed in bedrock or the overlying unconsolidated deposits. Wells drilled in bedrock typically are cased from the surface down to competent bedrock without grouting and consist of an open hole through the .

water-bearing zone, whereas those constructed in unconsolidated sand/gravel may be through open-ended casing or screened in the water-bearing zone.

Land use is predominately forested (56.1 percent), mostly on private land except for a small portion of Pennsylvania state forest, one state park, and several state game lands. Agricultural land use consists mainly of pasture/hay (23.5 percent) and cropland (11.4 percent; Figure 1B) (Multi-Resolution Land Characteristics Consortium, 2014). During the summer of 2016, Bradford County was second among all counties in the state for the most horizontal unconventional gas wells drilled (n = 1,339; Pennsylvania Department of Environmental Protection, 2017), and the related infrastructure (well pads, pipelines, and so forth) has become a relatively recent part of the landscape.

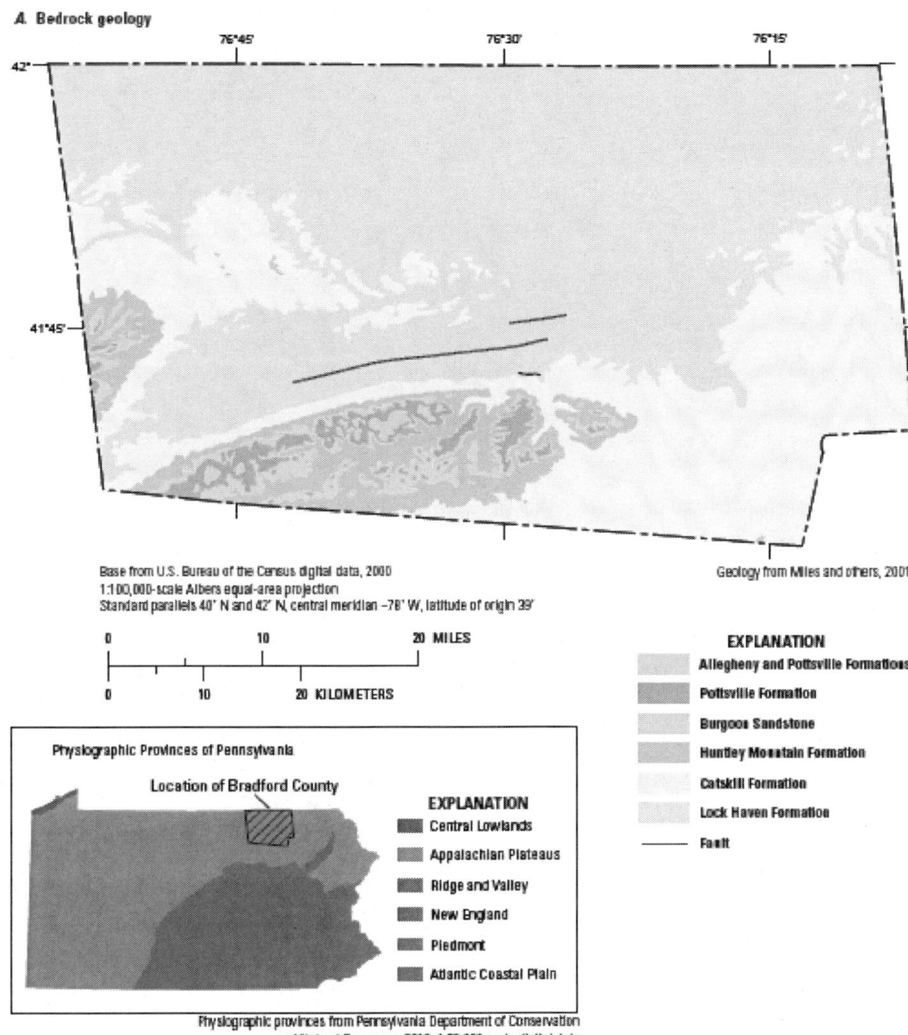

Figure 1. (Continued).

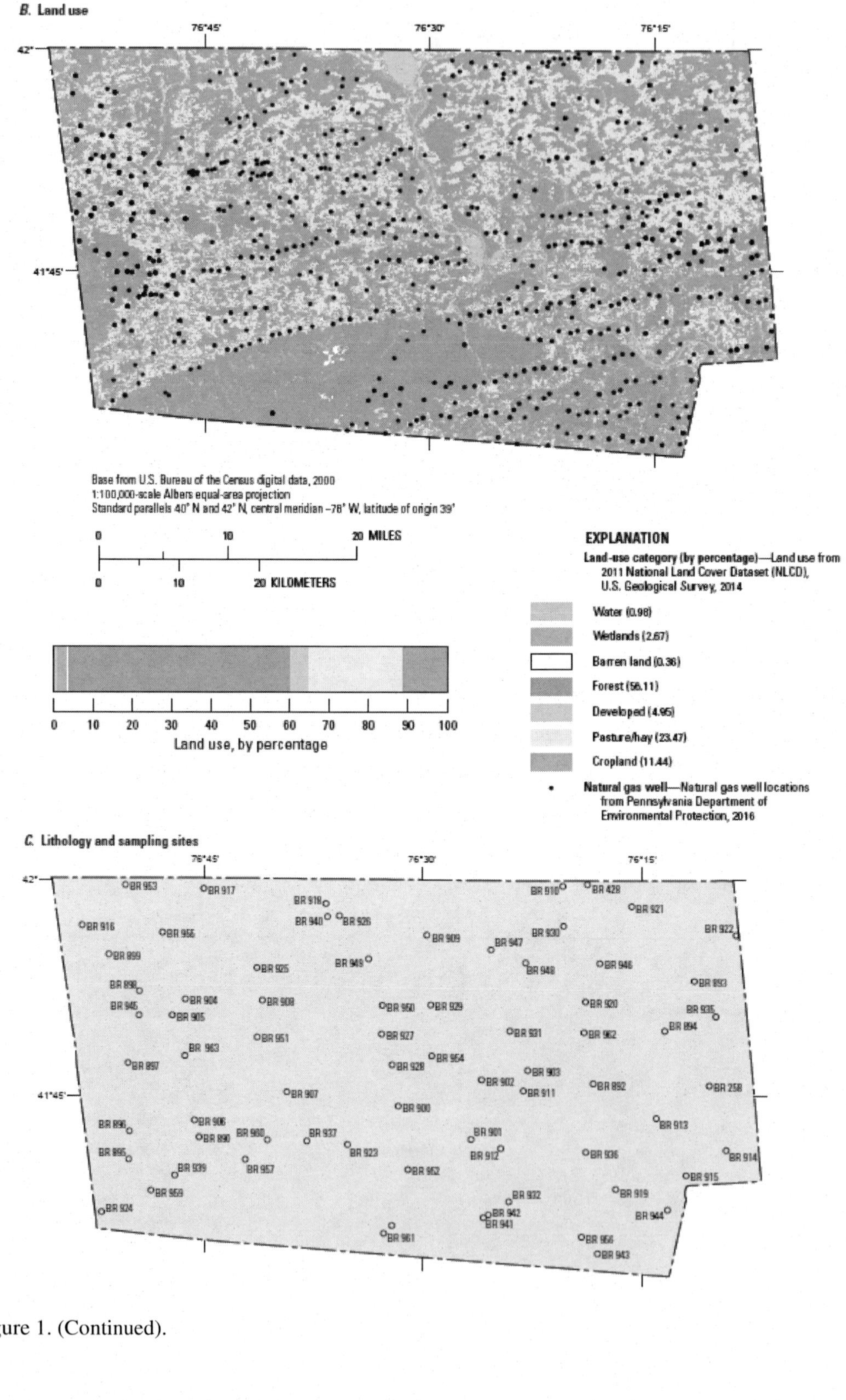

Figure 1. (Continued).

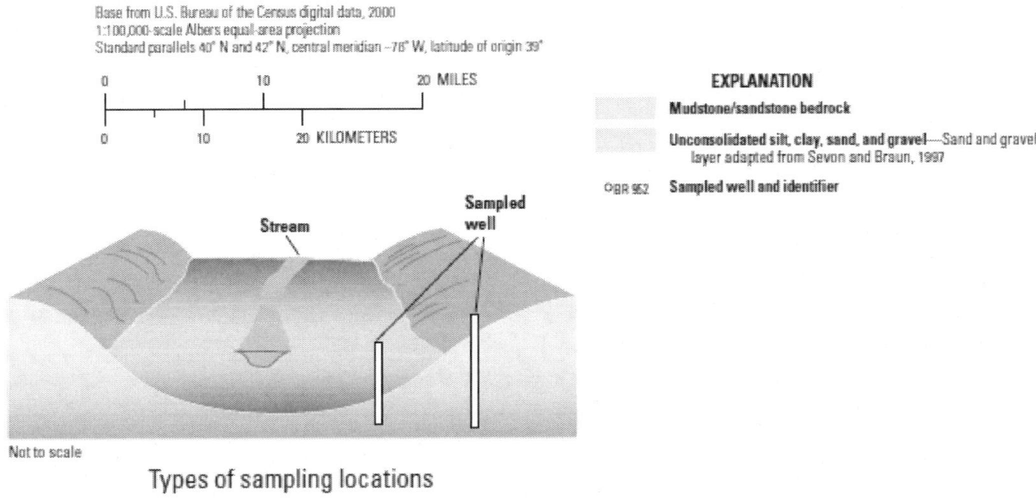

Figure 1. *A*, bedrock geology, *B*, land use, and *C*, lithology and sampling sites in Bradford County, north-central Pennsylvania. Bradford County is in a section of the glaciated region of the Appalachian Plateaus Physiographic Province.

Wetland and water resources compose 2.67 percent and 0.98 percent of the landscape, respectively. The remaining land area is developed/residential land (4.95 percent) with a largely rural population of 62,622 recorded in 2010 (U.S. Census Bureau, 2011). Some of the most densely populated areas in the county include the boroughs of Sayre, Athens, and Towanda, and the townships of Athens, Canton, and Ridgebury, with populations ranging from approximately 2,000 to 6,000 residents (U.S. Census Bureau, 2011).

Topography in Bradford County consists mostly of low to moderate relief with rounded hills and valleys with elevations ranging from 636 feet (ft) to 2,710 ft above sea level (Sevon, 2000; U.S. Geological Survey, 2009). The main streams within the county are Towanda, Sugar, Wysox, and Wyalusing Creeks that flow into the north branch of the Susquehanna River watershed, which eventually drains to the Chesapeake Bay. The local climate provides for a wet season of increased precipitation from April to June, with a mean annual rainfall of 37 inches per year (in/yr), of which approximately 10–12 inches recharges the aquifer (Risser and others, 2008). Groundwater levels fluctuate seasonally with increased recharge in the spring (March to May), a relatively stable water table during winter (November to February), and a steady decline during the summer (June to August) months owing to increased evapotranspiration (Figure 2). Average temperatures throughout the year in Bradford County range from about 33.6 to 55.6 degrees Fahrenheit (PRISM Group at Oregon State University, 2006a, b).

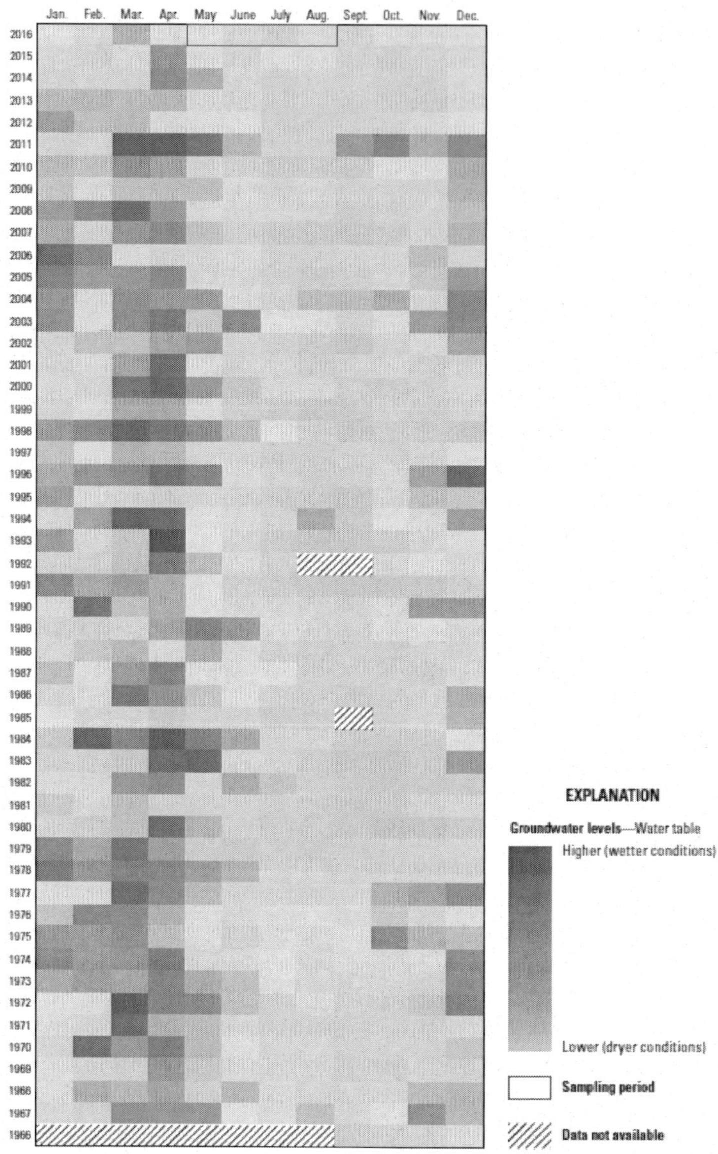

Figure 2. Groundwater levels for the period of record (1966–2016) for U.S. Geological Survey BR 92 Bradford County observation well (414330076280501). The water table in Bradford County fluctuates seasonally with increased precipitation/recharge (wetter conditions) in the spring (March to May) and declines during the summer (June to August) due to increased evapotranspiration (drier conditions). Groundwater samples were collected across Bradford County, Pennsylvania, from May to August 2016 (shown in red).

STUDY METHODS

To provide a characterization of groundwater quality in Bradford County, 72 private domestic-supply wells throughout the county (Figure 1) were selected for sampling during the summer months of 2016 (Figure 2). One sample from each well was analyzed for physical and chemical properties, including major ions, nutrients, trace elements, VOCs, ethylene and propylene glycol, gross-alpha/beta-particle activity, uranium, radon-222, and dissolved gases. A subset of samples was analyzed for radium isotopes (radium-226 and -228) and for the isotopic composition of methane. A list of wells sampled, ancillary data, and a National Water Information System (NWIS) website URL to water quality results used for analysis are provided by Clune and Cravotta (2018). Using the station identification numbers, the NWIS water quality results can also be retrieved with the dataRetrieval R package (Hirsh and DeCicco, 2015; Hirsh and others, 2015a, 2015b) or through user defined queries using USGS web services (U.S. Geological Survey, 2018). Additionally, quality assurance data results collected during sampling are available (Clune and Cravotta, 2018).

Selection of Sampling Locations

A computerized, stratified random site-selection approach was used to design a groundwater-quality network for sampling (Scott, 1990) by creating an equal-area grid of cells with a randomly placed population of potential sites across the county. The program arbitrarily specified one site from each cell as a primary sampling point and three additional sites as alternate sampling locations. The best available well record data within Bradford County were obtained from the Pennsylvania Groundwater Information System database (Pennsylvania Bureau of Topographic and Geologic Survey, 2014) and from local well drillers. Potential wells for sampling were selected from these data based on the criteria that the well was (1) used for domestic or stock purposes, (2) had an associated well drillers record, (3) was drilled after 1970, (4) was not hand dug, and (5) the top of the casing was above land surface. Wells located within 1 mile of one of the computer-generated random sampling points were identified, and well owners were contacted for permission to sample and to schedule a plumbing review to verify an untreated sample representative of the aquifer could be collected. Private domestic-supply wells meeting the criteria and well verification were prioritized for sampling.

Collection and Analysis of Samples

Sampling sites included 72 groundwater well locations throughout Bradford County (Figure 1*C*) drilled with an open hole interval usually in bedrock and sometimes unconsolidated deposits. Groundwater samples were collected from May through August 2016 (Figure 2) following protocols outlined in the USGS National Field Manual (U.S. Geological Survey, variously dated). Groundwater samples were collected using Teflon tubing attached to a sampling point (outside spigot, pressure tank, and so forth) prior to any water treatment. Wells were purged until field parameters (pH, specific conductance, water temperature, turbidity, and dissolved oxygen) stabilized and the final readings were recorded. Samples were then collected for laboratory analysis.

The samples for total dissolved solids, major ion, trace metals, and nutrient chemistry were filtered onsite through a disposable filter with a 0.45-micrometer pore size and analyzed by the National Water Quality Lab in Denver, Colorado. Samples for major ions were preserved with nitric acid if the pH was greater than 2 and analyzed at the lab by inductively coupled plasma atomic emission spectrometry, ion chromatography, or automated colorimetry (EPA Method 524). Nutrient analyses were performed by automated colorimetry with persulfate digestion when required (Fishman, 1993; Patton and Kryskalla, 2011).

Radiological samples for alpha/beta particles were preserved with nitric acid if pH was greater than 2 and analyzed at ALS Laboratories in Fort Collins, Colorado, by gas flow proportional counting (EPA Method 900.0). A subset of samples that had gross alpha greater than 5 picocuries per liter (pCi/L) were analyzed further for radium-226 and -228 by radon emanation, chemical separation, and gas flow proportional counting (EPA Method 903.1 and 904.0). Radon-222 samples were obtained through an inline septum with a gastight syringe to avoid atmospheric contact and analyzed at the lab by liquid scintillation (ASTM Method D 5072-16).

Dissolved gases sampling used a collection bag, and samples were analyzed at Isotech Laboratories, Champaign, Illinois, by gas compositional analysis to determine the concentration of fixed gases and light hydrocarbon gases dissolved in the sample (Dai and others, 2012). Samples containing a sufficient concentration of methane were further analyzed for stable carbon isotopes ^{12}C and ^{13}C and the stable hydrogen isotopes ^{1}H and ^{2}H (deuterium). The $^{13}C/^{12}C$ and $^{2}H/^{1}H$ ratios were determined by an isotope ratio mass spectrometric analysis that compared the sample to a reference standard. The carbon isotope ratio value in methane ($\delta^{13}C$ CH_4) is reported in terms of per mil (‰) notation with respect to the Vienna Peedee belemnite standard. The hydrogen isotope ratio value in methane (δD CH_4) is reported in terms of per mil notation with respect to the Vienna Standard Mean Ocean Water standard.

The remaining unfiltered samples were analyzed by Seewald Laboratories, Inc., in Williamsport, Pennsylvania. VOCs, dissolved gases, and glycol samples were collected onsite using a stainless-steel manifold fitting to fill vials to reduce turbulence. VOC vials were preserved with hydrochloric acid and analyzed by purge and trap gas chromatography/mass spectrometry (EPA Method 524.2). Dissolved gas samples for ethane and propane were determined by analyzing a portion of the headspace with a gas chromatograph equipped with a flame ionization detector (PA DEP Method 3686). Based upon the equilibrium gas concentration, a liquid sample concentration of these gases is calculated. Samples for glycols and alcohols were evaluated using the direct injection method, which uses a gas chromatograph equipped with a flame ionization detector that is chromatographically optimized for the separation and determination of the compounds of interest (EPA Method 8015 D). Oil and grease samples were preserved with hydrochloric acid and analyzed in the lab using hexane to determine the petroleum hydrocarbon fraction (EPA Method 1664 B). After sterilizing and flaming the sampling point, the bacteria sample was collected and analyzed by the lab using the Colilert color method in which samples were enumerated to give counts of total coliform and *Escherichia coli* (Standard Method 9223).

Graphical and Statistical Analyses

Various graphical and statistical techniques were used in this report to compare water-quality data among different sites, to distinguish natural and anthropogenic sources of dissolved constituents, and to identify possible factors affecting the occurrence or transport of solutes in the aquifers in the study area. Techniques included bivariate scatter plots, mass ratios of chloride to bromide (Cl/Br), trilinear diagrams, nonparametric statistical approaches, correlation coefficients (Spearman's rho), principal components analysis (PCA), and aqueous speciation computations. The boxplots in the appendix considered pH (Figure 4.1), specific conductance (Figure 4.2), redox (Figure 4.3), bedrock formation (Figure 4.4), and topographic position index (Figure 4.5) to classify the data. The boxplots show a notched interval around the median that can be used by a reader to judge the significance of potential differences (Velleman and Hoaglin, 1981; Helsel and Hirsch, 2002). If the notched intervals around the medians for sample subsets do not overlap, the medians are statistically different at the 95-percent confidence interval. An extensive description of the graphical and statistical methods used in this report and the basis for classifications used in the appendix are described in Gross and Cravotta (2017).

Quality Assurance and Quality Control

For quality control (QC), filtered and unfiltered replicate samples were collected from wells BR 907, BR 910, BR 913, BR 939, and field blanks were processed onsite for BR 921, BR 923, and BR 940. All these samples were submitted to the laboratories for analysis. These results demonstrate that, for most constituents, field equipment or sampling methodology did not contaminate the sample. The results of the equipment blank indicate that sampling equipment may introduce low levels of two constituents into the samples at concentrations of as much as about 0.22 micrograms per liter (μg/L) for ammonia and 0.05 μg/L for molybdenum. The results of one of the three field blanks (BR 921) indicate that sampling methodology may introduce low levels of four constituents into samples at concentrations of as much as about 0.05 milligrams per liter (mg/L) for calcium, 0.10 mg/L for sodium, 0.04 mg/L for chloride, and 0.03 mg/L for silica.

Results from three sets of replicate samples indicate combined sampling and analytical precision (reproducibility) was within 5 percent and (or) concentrations were less than two times the minimum reporting level for most constituents. Within two times the minimum reporting level, the concentrations were less precise but still within 20 percent, except for gross alpha and beta radioactivity. For the replicate sample collected from BR 939, calcium values were 77.3 and 66.6 mg/L with a difference of 10.7 mg/L (15 percent), and methane had low detected values of 0.0042 and 0.0037 mg/L with a difference of 0.0005 mg/L (13 percent). For the replicate sample collected from BR 910, bromide was 0.440 and 0.493 mg/L with a difference of 0.053 mg/L (11 percent difference). For the replicate sample at BR 907, ammonia was 0.29 and 0.33 mg/L with a difference of 0.04 (14 percent), and radon 222 was 324 and 349 pCi/L with a 25-pCi/L difference (7 percent). The two other replicate samples with values for calcium, methane, bromide, ammonia, and radon 222 had a less than 3-percent difference or were less than two times the minimum reporting level.

For quality assurance (QA) of the inorganic chemical analyses, intrasample characteristics were evaluated using standard procedures described by Hem (1985) and Fishman and Friedman (1989). Evaluations of accuracy and precision included comparison of field- and laboratory-measured values for pH and specific conductance for each sample. Additional QA/QC checks involved comparisons of the computed cation and anion equivalents concentrations and the corresponding ionic charge balance, comparisons of the ratios of cation or anion equivalents to specific conductance, and comparisons of total dissolved solids (TDS) computed as the sum of major ion concentrations to the measured residue on evaporation (ROE) at 180 degrees Celsius (°C). Lastly, the measured specific conductance was compared to the computed specific conductance, which was estimated as the sum of ionic conductivities after accounting for aqueous speciation (McCleskey and others, 2012). The values of specific conductance measured in the field and laboratory were consistent with one another and with the computed values of specific conductance and TDS on the basis of measured solute concentrations. The measured TDS concentration as ROE by two different laboratories and the computed TDS as the sum of constituent concentrations were in close agreement.

Groundwater Quality and Comparison to Drinking Water Health Standards

Analytical results for the 72 groundwater samples collected in Bradford County are provided in the following sections and compared to EPA drinking-water standards and health advisories (U.S. Environmental Protection Agency, 2012a). The EPA has established maximum contaminant levels (MCLs), secondary maximum contaminant levels (SMCLs), and action levels for some constituents in drinking water (Table 1). MCLs generally are set because elevated concentrations of these constituents may cause adverse health effects. SMCLs generally are set for aesthetic reasons; elevated concentrations of these constituents may impart an undesirable taste or odor to water. These MCLs and SMCLs may be used as a guideline for private well owners but are only regulated for public drinking-water supplies.

Physical and Chemical Properties

Physical and chemical properties discussed in this section include temperature, pH, specific conductance, dissolved oxygen, and turbidity. Most of these properties are unstable and are determined in the field at the time a water sample is collected.

The temperature of the groundwater samples ranged from 9.4 to 18.0°C, with a median of 11.5°C. These temperatures generally were less than the daytime air temperatures during sampling, which reflects the generally cool groundwater environment. Dissolved gases and carbonate minerals can dissolve to a greater extent in cooler water than in warmer water (Hem, 1985).

Dissolved oxygen (DO) concentrations ranged from 0.03 to 10.6 mg/L; the median concentration was 1.21 mg/L (Table 1). Generally, most of the groundwater had DO concentrations that were substantially less than saturation at the sample temperature, indicating the waters had been out of contact with the atmosphere. Low DO concentrations are associated with higher pH values; 21 of the 30 water samples with a DO concentration of less than 1 mg/L had a pH greater than 7.5. Low DO concentrations are related to chemical or biochemical reactions that consume oxygen and may be associated with chemically reducing conditions that promote the release of iron, manganese, and associated metals from rock and other solids in contact with water along its flow path. The chemical reactions that consume oxygen generally involve organic carbon that can be naturally occurring in soil or aquifer materials, or can be introduced from industrial, agricultural, or domestic wastes. Of the 72 well samples in this study, 36.1 percent were classified as "anoxic" (DO $<$ 0.5 mg/L), 50.0 percent as "oxic" (DO $\geq$ 0.5 mg/L), and 13.9 percent as "mixed" (DO $>$ 0.5 mg/L with manganese (Mn) $>$ 0.05 mg/L or iron (Fe) $>$ 0.1 mg/L), based on criteria of McMahon and Chapelle (2008) (Figure 4.3). Of the 26 samples classified as anoxic, 12 had chemical characteristics consistent with MnIV and FeIII reducing conditions as defined by McMahon and Chapelle (2008) (DO $<$ 0.5 mg/L, nitrate (NO_3) $<$ 0.5 mg/L, Mn$>$0.05 mg/L, Fe $>$ 0.1 mg/L, and sulfate (SO_4) $>$ 0.5 mg/L), and two were identified with more strongly reducing conditions necessary for methanogenesis (DO $<$ 0.5 mg/L; $NO_3 <$ 0.5 mg/L, Mn $>$ 0.05 mg/L, Fe $>$ 0.1 mg/L, and $SO_4 <$ 0.5 mg/L). Both samples identified as methanogenic had elevated concentrations of methane; BR 910 and BR 921 had methane concentrations of 77 mg/L and 26 mg/L, respectively.

The pH is a measure of acidity and is related to the potential corrosivity of the water and its potential to leach metals, such as lead and copper, from pipes and plumbing. The standard pH scale is 0–14, with 7 being neutral. Generally, pH values of 6.5 to 7.5 are considered near neutral, values less than 6.5 are considered acidic, and values greater than 7.5 are considered basic or alkaline. The field pH of water samples collected in Bradford County ranged from 6.18 to 9.31; the median pH was 7.47. The pH of 6 of the 72 samples (8.3 percent) was outside the EPA SMCL range of 6.5–8.5 (U.S. Environmental Protection Agency, 2012a). Two samples had a pH less than 6.5, and four samples had a pH greater than 8.5 (Table 1 and Figure 3*C*). Generally, the pH varied widely within and among the three bedrock aquifers sampled; median pH values were greatest for the Lock Haven Formation (7.6), least for Huntley Mountain Formation (6.8), and intermediate for the Catskill Formation (7.4) (Figure 4.4).

Specific conductance (SC) is a gross measure of the ability of ions in water to conduct an electrical current and is expressed in units of microsiemens per centimeter (μS/cm). The higher the value for specific conductance, the higher the concentrations of total dissolved solids and associated ions in the water. The field-measured SC for the 72 groundwater samples ranged from 72.4 to 3,410 μS/cm at 25°C; the median specific conductance was 365 μS/cm (Table 1). Values were similar for the laboratory-measured SC (Table 1), which is useful because the laboratory measured SC avoids potential issues with air bubbles forming on the electrode surfaces when the sample is first drawn from the subsurface.

Turbidity is a measure of the amount of solid particles suspended in water that block the transmission of light through the sample. Turbidity is expressed in nephelometric turbidity ratio units (NTRU), which quantify the degree to which light is scattered by solid particles suspended in water. The higher the NTRU, the more turbid the water. Turbidity concentrations ranged from 0.12 to 247 NTRU; the median concentration was 1.19 NTRU (table 1). In general, samples with high turbidity could be expected to have total concentrations of constituents that include contributions from suspended particles, in addition to the dissolved concentration.

Major Ions

Major ions are frequently derived from the dissolution of common minerals, including carbonates, silicates, oxides, sulfates, and sulfides, and can be influenced by ion-exchange, redox processes, and mixing of freshwater with residual brines that remain in the aquifer matrix or that could be mobilized from deep sources. The concentrations of major ions, TDS, salinity, and SC of groundwater are directly related, and generally expected to increase with, progressive evaporation or dissolution of minerals (Hem, 1985). The concentrations of trace elements in solution may increase with TDS or SC, not only because of the release of trace constituents with the major ions dissolved from minerals, but because of the potential for displacement of adsorbed or exchangeable trace ions from mineral surfaces by the major ions.

Major cations (positively charged ions such as calcium, magnesium, sodium, and potassium), major anions (negatively charged ions such as sulfate, chloride, fluoride, and bicarbonate), and nonionic solutes (uncharged solutes such as silica) typically are present in natural waters at concentrations greater than 1 mg/L, whereas dissolved trace constituents (such as iron, manganese, zinc, lead, copper, nickel, vanadium, molybdenum, arsenic, selenium, radium, uranium, lithium, and bromide) typically are present at concentrations less than 1 mg/L (Hem, 1985). Concentrations of DO and nutrients (such as nitrogen and phosphorus compounds) commonly range from values less than 1 mg/L to values greater than 1 mg/L because of biological processes that involve the production or metabolism of organic carbon compounds that contain nitrogen and phosphorus. Although biological (biochemical) processes can affect the concentrations of nutrients and trace constituents in groundwater directly or indirectly because of changes to pH and redox, such processes generally have minor effects on major ion concentrations.

Analyses for major ions were conducted on filtered samples to represent dissolved concentrations. Major anions, in order of decreasing median dissolved concentration, were bicarbonate (alkalinity), sulfate, chloride, fluoride, and bromide; major cations were calcium, sodium, magnesium, and potassium. These cations and anions, plus uncharged silica, are the primary contributors to TDS and SC. Of the major ions, only fluoride has an established EPA MCL in drinking water (4 mg/L) or SMCL (2 mg/L). Fluoride concentrations ranged from 0.03 to 0.78 mg/L, and none of the samples exceeded the EPA MCL or SMCL (Table 1).

The concentration of TDS can be measured by evaporating a given volume of water and weighing the ROE or by summing the measured concentrations of the major ions in a filtered sample (TDScalc). The measured TDS concentrations ranged from 52.7 to 2,030 mg/L; the median concentration was 218 mg/L. The TDScalc was comparable in value to the ROE and ranged from 42.0 to 1,940 mg/L. Out of 72 samples, 4 samples (5.6 percent) had TDS (ROE and TDScalc) that exceeded 500 mg/L, which is the EPA SMCL for TDS in drinking water. One of these four samples had elevated concentrations of sulfate (BR 946) that may result from the oxidation of pyrite, whereas three samples (BR 913, BR 921, and BR 927) had elevated concentrations of sodium and chloride. Elevated sodium and chloride concentrations could be caused by connate water in the formations and a zone of restricted groundwater flow that limits flushing by fresh recharge; mixing of saline and freshwater; or by the introduction of mineral salts from near-surface contaminant sources such as road deicing compounds, water softener additives, effluent from septic systems, or animal waste (Mullaney and others, 2009).

Table 1. Minimum, median, and maximum values of selected characteristics and constituents in groundwater samples collected in 72 domestic wells in Bradford County, Pennsylvania, May–August 2016. Available U.S. Environmental Protection Agency Maximum Contaminant Levels, Action Levels, and Secondary Maximum Contaminant Levels for analyzed constituents (2012 Edition of the Drinking Water Standards and Health Advisories [EPA 822-S-12-001])

Constituent (units)	n	Minimum	Median	Maximum	Results above the reporting level		Results exceeding standard			EPA MCL	EPA SMCL
					Number	Percent	Number	Percent	Graph		
Well characteristics											
Well depth (feet)	72	42	183	600	—	—	—	—	—	—	—
Physical properties											
Water temperature (°C)	72	9.40	11.5	18.0	72	100	—	—	—	—	—
Dissolved oxygen (mg/L)	72	0.03	1.21	10.6	72	100	—	—	—	—	—
Specific conductance, field (µS/cm)	72	72.4	365	3,410	72	100	—	—	—	—	—
pH, field (standard units)	72	6.18	7.47	9.31	72	100	6	8.3		—	6.5–8.5
Redox potential (mV)	71	−190	140	370	71	100	—	—	—	—	—
Turbidity (Nephelometric Turbidity Ratio Units)	70	0.12	1.19	247	70	100	—	—	—	—	—
Laboratory analysis											
Specific conductance, lab (µS/cm)	72	79.3	369	3,800	72	100	—	—	—	—	—
pH, lab (standard units)	72	6.96	8.09	9.16	72	100	3	4.2		—	6.5–8.5
Alkalinity, lab (mg/L $CaCO_3$)	72	26.1	178	347	72	100	—	—	—	—	—
Total dissolved solids (mg/L)	72	52.7	218	2,030	72	100	4	5.6		—	500
Suspended solids (mg/L)	72	15.0	15.0	488	5	6.94	—	—	—	—	—
Hardness, total (mg/L $CaCO_3$)	71	1.32	134	627	71	100	—	—	—	—	—
Major ions											
Calcium (mg/L)	72	0.04	39.6	164	72	100	—	—	—	—	—
Magnesium (mg/L)	72	0.01	8.37	51.6	72	100	—	—	—	—	—
Sodium (mg/L)	72	2.18	19.1	760	72	100	35	48.6		—	30–60
Potassium (mg/L)	72	0.22	1.40	4.89	72	100	—	—	—	—	—
Chloride (mg/L)	72	0.77	5.28	1,020	72	100	1	1.4		—	250
Fluoride (mg/L)	72	0.03	0.12	0.78	71	98.6	0	0		4	2
Sulfate (mg/L)	72	0.06	13.5	418	71	98.6	0	0		—	250
Silica (mg/L as SiO_2)	72	6.05	11.1	15.9	72	100	—	—	—	—	—
Trace elements											
Aluminum (µg/L)	72	< 3.00	< 3.00	771	10	13.9	1	1.4		—	50
Antimony (µg/L)	72	< 0.02	< 0.02	0.30	31	43.1	0	0		6	—
Arsenic (µg/L)	72	< 0.05	0.63	23.2	67	93.1	2	2.8		10	—
Barium (µg/L)	72	< 0.25	142	4,160	71	98.6	2	2.8		2,000	—
Beryllium (µg/L)	72	< 0.01	0.02	0.10	8	11.1	0	0		4	—
Boron (µg/L)	72	< 5.00	68.9	567	70	97.2	—	—	—	—	—
Bromide (mg/L)	72	< 0.01	0.02	8.60	58	80.6	—	—	—	—	—
Cadmium (µg/L)	72	< 0.03	0.03	0.31	2	2.78	0	0		5	—
Chromium (µg/L)	72	< 0.30	< 0.30	5.00	2	2.78	0	0		100	—
Cobalt (µg/L)	72	< 0.03	0.05	0.75	17	23.6	—	—	—	—	—
Copper (µg/L)	72	< 0.20	0.82	82.3	37	51.4	0	0		—	1,000
Iron (µg/L)	72	< 4.00	11.34	1,830	44	61.1	8	11.1		—	300
Lead (µg/L)	72	< 0.02	0.10	2.61	53	73.6	0	0		15	—
Lithium (µg/L)	72	0.41	21.8	1,510	72	100	—	—	—	—	—
Manganese (µg/L)	72	< 0.40	10.5	1,120	56	77.8	22	30.6		—	50
Molybdenum (µg/L)	72	< 0.05	0.48	2.42	68	94.4	—	—	—	—	—
Nickel (µg/L)	72	< 0.20	< 0.20	2.00	30	41.7	—	—	—	—	—

Constituent (units)	n	Minimum	Median	Maximum	Results above the reporting level		Results exceeding standard			EPA MCL	EPA SMCL
					Number	Percent	Number	Percent	Graph		
Trace elements—Continued											
Selenium (µg/L)	72	< 0.05	< 0.05	0.56	30	41.7	0	0		50	—
Silver (µg/L)	72	< 0.02	< 0.02	10.0	1	1.39	0	0		—	100
Strontium (µg/L)	72	< 0.80	539	3,860	71	98.6	—	—	—	—	—
Zinc (µg/L)	72	< 2.00	3.23	981	47	65.3	0	0		—	5,000
Nutrients											
Kjeldahl nitrogen, total (mg/L as Nitrogen)	72	< 0.07	0.07	1.71	33	45.8	—	—	—	—	—
Ammonia (mg/L as Nitrogen)	72	< 0.01	< 0.01	1.60	42	58.3	—	—	—	—	—
Nitrite (mg/L as Nitrogen)	72	< 0.001	< 0.001	0.23	17	23.6	0	0		1	—
Nitrate + Nitrite (mg/L as Nitrogen)	72	< 0.04	< 0.04	7.01	36	50.0	0	0		10	—
Orthophosphate (mg/L as Phosphorus)	72	< 0.004	0.01	0.16	63	87.5	—	—	—	—	—
Bacteria											
Total coliform (MPN/100 mL)	71	< 1	< 1	> 2,420	35	49.3	35	49.3		0	—
Escherichia Coli (MPN/100 mL)	71	< 1	< 1	> 2,420	6	8.5	6	8.5		0	—
Radiochemicals											
Gross alpha radioactivity, 30-d (pCi/L)	72	−0.89	1.00	8.3	39	54.2	0	0.0		15	—
Gross alpha radioactivity, 72-hr (pCi/L)	72	−0.33	1.70	11.2	52	72.2	0	0.0		15	—
Gross beta radioactivity, 30-d (pCi/L)	72	−0.50	2.34	8.1	57	79.2	12	16.7		4	—
Gross beta radioactivity, 72-hr (pCi/L)	72	0.03	2.40	5.89	64	88.9	12	16.7		4	—
Radon-222 (pCi/L)	71	34.4	614	8,540	71	100	50	70.4		[1]300	—
Uranium (pCi/L)	72	< 0.01	0.26	10.4	62	86.1	0	0.0		30	—
Radium-226 (pCi/L)	8	0.16	1.08	2.08	8	100	0	0.0		5	—
Radium-228 (pCi/L)	8	−0.02	0.31	1.54	4	50.0	0	0.0		5	—
Glycols, alcohols and oil/grease											
Ethanol (mg/L)	72	< 5.00	< 5.00	< 5.00	0	0	—	—	—	—	—
Ethylene glycol (mg/L)	72	< 5.00	< 5.00	< 5.00	0	0	0	0.0		14	—
Isobutyl alcohol (mg/L)	72	< 5.00	< 5.00	< 5.00	0	0	—	—	—	—	—
Isopropyl alcohol (mg/L)	72	< 5.00	< 5.00	< 5.00	0	0	—	—	—	—	—
Methanol (mg/L)	72	< 5.00	< 5.00	< 5.00	0	0	—	—	—	—	—
n-Butanol (mg/L)	72	< 5.00	< 5.00	< 5.00	0	0	—	—	—	—	—
n-Propanol (1-Propanol) (mg/L)	72	< 5.00	< 5.00	< 5.00	0	0	—	—	—	—	—
Propylene glycol (mg/L)	72	< 5.00	< 5.00	< 5.00	0	0	—	—	—	—	—
Oil and grease (mg/L)	72	< 5.00	< 5.40	< 6.00	0	0	—	—	—	—	—
Volatile organic compounds[2]											
Bromodichloromethane (CHBrCl2) (µg/L)	72	< 0.50	< 0.50	1.04	1	1.39	0	0		600	—
Dibromochloromethane (µg/L)	72	< 0.50	< 0.50	1.58	1	1.39	0	0		60	—
Tribromomethane (µg/L)	72	< 0.50	< 0.50	0.81	1	1.39	0	0		80	—
Dissolved gases											
Methane (mg/L)	72	< 0.0002	0.006	77	64	88.9	9	12.5		[3]> 7	—
Ethane (mg/L)	72	< 0.0002	0.0002	0.13	15	20.8	—	—	—	—	—
Propane (mg/L)	72	< 0.0002	0.0002	0.001	2	2.78	—	—	—	—	—
$\delta^{13}C$ (per mil in CH_4)	6	−65.05	−44.37	−38.23	6	100	—	—	—	—	—
δD (per mil in CH_4)	6	−257.2	−203.65	−191.2	6	100	—	—	—	—	—

EXPLANATION

Exceeds MCL | Exceeds SMCL | No exceedance

[n, number of results; EPA, U.S. Environmental Protection Agency; MCL, Maximum Contaminant Level; SMCL, Secondary Maximum Contaminant Level; —, no MCL, Action Level, or SMCL established;°C, degrees Celsius; mg/L, milligrams per liter; µS/cm, microsiemens per centimeter; mV, millivolts; CaCO3, calcium carbonate; SiO2, silicon dioxide; µg/L, micrograms per liter; <, less than; MPN/100 mL, most probable number per 100 millimeters; >, greater than; pCi/L, picocuries per liter; 72-hr, sample analyzed for gross alpha concentration at approximately 72 hours after sample collection as referenced to a detector calibrated using thorium-230 (230Th); 30-d, sample used for the 72-hour gross alpha analysis is counted a second time approximately 30 days after the initial count as referenced to a detector calibrated using 230Th; δ, delta; mil, million].

[1]Proposed EPA MCL for states without an enhanced indoor air program.

[2]Volatile organic compounds (VOCs) with detections are shown. A full listing of all VOCs sampled but not detected are shown in Appendix 4.

[3]Pennsylvania action level for methane.

Dissolved concentrations of sodium ranged from 2.18 to 760 mg/L; the median concentration was 19.1 mg/L.

The EPA has issued a taste threshold for sodium in public water supplies that recommends reducing sodium concentrations to between 30 and 60 mg/L (U.S. Environmental Protection Agency, 2003). This recommendation is a guideline and is not federally enforceable. Salt used for food, road deicing, and other applications is mainly composed of sodium chloride. Thus, high concentrations of sodium and chloride may result from the addition of salt to groundwater and may be related to land use as well as the bedrock and (or) unconsolidated lithology. Dissolved concentrations of chloride ranged from 0.77 to 1,020 mg/L; the median concentration was 5.28 mg/L (Table 1, Figure 3*B*). Out of the 72 samples, only 1 sample (1.4 percent), BR 913, exceeded the EPA SMCL of 250 mg/L for chloride in drinking water.

The alkalinity of a solution indicates its capacity to neutralize acid and commonly results from dissolved carbonate and bicarbonate ions (Hem, 1985). Alkalinity, also referred to as the acid neutralizing capacity, generally increases with the pH of a water sample. Alkalinity ranged from 26.1 to 347 mg/L as calcium carbonate ($CaCO_3$); the median concentration was 178 mg/L (Table 1). The highest alkalinities were from the Lock Haven Formation, and the lowest were from Huntley Mountain Formation (Figure 4.4).

The hardness of the 72 well-water samples ranged from 1.32 to 627 mg/L as $CaCO_3$, with a median value of 134 mg/L as $CaCO_3$. Hardness is directly related to the concentrations of calcium and magnesium (computed as the sum of calcium, in mg/L, multiplied by a factor of 2.5 plus magnesium, in mg/L, multiplied by a factor of 4.1) and generally is comparable to the alkalinity. The alkalinity and associated cations can be released into groundwater by the dissolution of calcium- and magnesium-bearing minerals, such as calcite ($CaCO_3$) and dolomite ($CaMg(CO_3)_2$), that generally are associated with limestone and other calcareous sedimentary rocks. Hard water decreases the lathering of soap and increases accumulation of mineral deposits in plumbing and cooking utensils. Using a common hardness classification (Durfor and Becker, 1964), the measured values indicate that 8 (11.1 percent) of the 72 water samples are soft (less than 60 mg/L as $CaCO_3$), 19 samples (26.4 percent) are moderately hard (61 to 120 mg/L as $CaCO_3$), 27 samples (37.5 percent) are hard (121 to 180 mg/L as $CaCO_3$), and 18 samples (25.0 percent) are classified as very hard (greater than 180 mg/L as $CaCO_3$). Median hardness values were 52.1, 128, and 147 mg/L as $CaCO_3$ for the Huntley Mountain, Catskill, and Lock Haven Formations, respectively. Hardness varied with pH, with hardness generally greatest at near-neutral pH values (6.5 to 7.5). Water samples with low pH (less than 6.5) and very high pH (greater than 8.0) typically had hardness less than 60 mg/L and were considered soft. The high-pH waters with low hardness typically have elevated sodium, which may result from natural water-softening processes, which are described later. There are no health-related standards established specifically for hardness in drinking water.

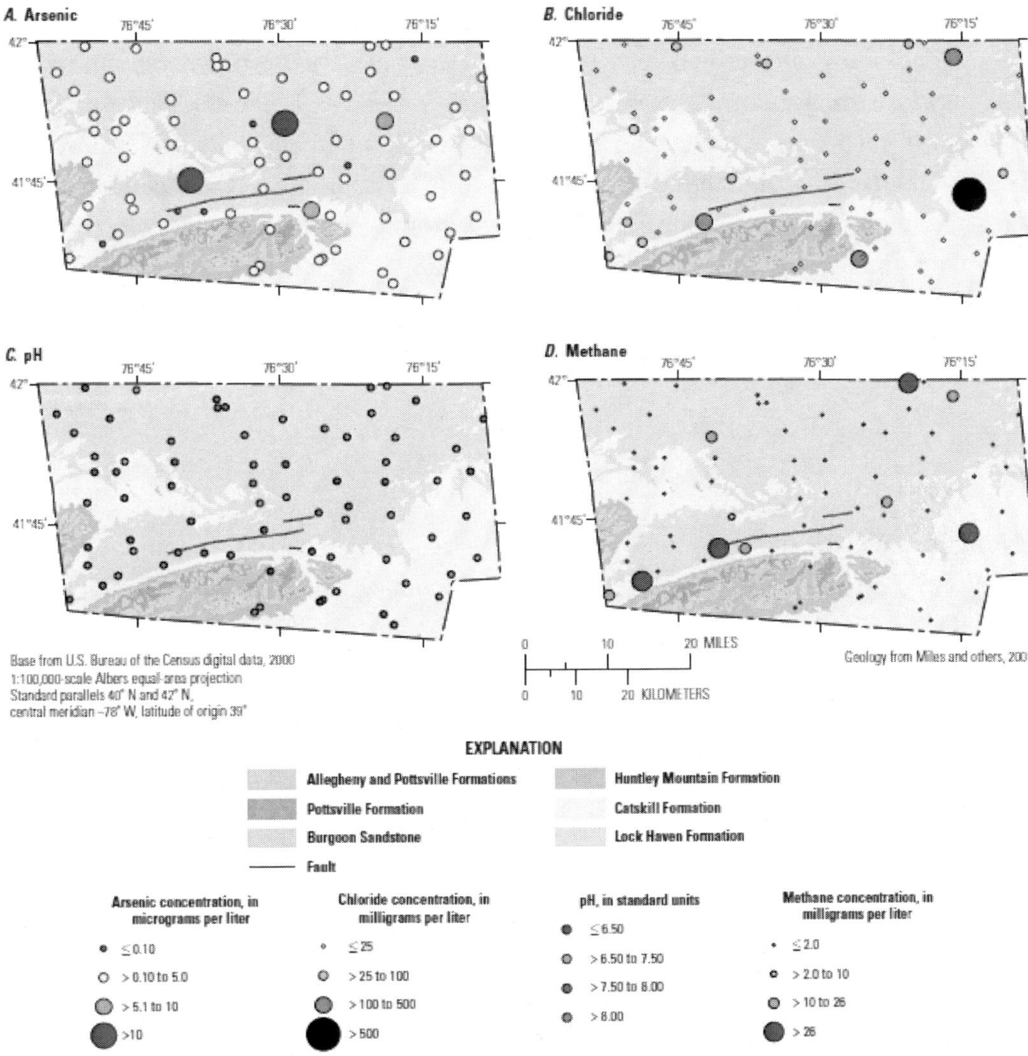

Figure 3. Spatial distribution of *A*, arsenic concentrations, *B*, chloride concentrations, *C*, pH values, and *D*, methane concentrations in 72 wells sampled and bedrock geology in Bradford County, Pennsylvania. (≤, less than or equal to; >, greater than).

Hard water has greater potential than soft water to form scale or encrustations on plumbing and fixtures. Water resources engineers commonly identify the potential for encrustation or corrosion on the basis of the Langelier Index (LI), which provides an indication for the potential for lead and copper to enter water supplies from pipes and plumbing (Snoeyink and Jenkins, 1981). The LI, which is the difference between the measured pH and the pH at equilibrium with calcium carbonate ($CaCO_3$), is equal in value to the calcite saturation index (SICAL), evaluated in more detail later. If the LI or

SICAL is positive, the pH is greater than that at equilibrium with $CaCO_3$ and the water will tend to deposit $CaCO_3$ coatings or scale that can insulate pipes, boilers, and other components of a system from contact with water; however, if negative, then the water is undersaturated with $CaCO_3$ and will tend to be corrosive in the distribution system. An optimum condition is desired whereby the LI or SICAL is close to zero, whereby the water will neither be strongly corrosive nor scale forming. For the 72 groundwater samples, SICAL ranged from −2.6 to 0.8. Of the 72 samples, 11 (2.8 percent) had SICAL values that were less than −0.5, indicating potentially strongly corrosive characteristics; the remaining samples would be considered neither strongly corrosive nor scale forming. Two samples characterized as strongly corrosive, BR 893 and BR 952, had SICAL less than −1.0 and hardness less than 30 mg/L as $CaCO_3$, but pH values were 8.1 and 7.0, respectively.

Nutrients

Nutrients include nitrogen and phosphorous species. Nitrogen is found in water principally as nitrate (NO_3), nitrite (NO_2), and ammonia (NH_4), whereas phosphorus is present mainly as orthophosphate (PO_4). Nutrients are essential for plant growth; however, the presence of elevated nutrient concentrations generally is an indicator of anthropogenic sources, which might include fertilizers, storm runoff, animal wastes, and effluent from septic systems. Summary statistics for nutrients are given in Table 1. Nitrate is the most prevalent nitrogen species in oxic groundwater, but all nitrate concentrations were still less than the EPA MCL of 10 mg/L for nitrate as N (Table 1). Excessive nitrate in drinking water causes a health risk, especially in infants, because it disrupts oxygen flow in blood. Infants below the age of six months who drink water containing nitrate in excess of the MCL could become seriously ill, develop symptoms including shortness of breath, and blue-baby syndrome. Because concentrations of nitrite are so low, nitrate plus nitrite concentrations essentially represent nitrate concentrations. The concentration of nitrate in samples ranged from less than 0.04 to 7.01 mg/L as N; the median concentration was less than 0.04 mg/L as N. The concentration of orthophosphate ranged from less than 0.004 to 0.16 mg/L with a median of 0.01 mg/L.

Bacteria

All samples were analyzed for total coliform and *Escherichia coli* (*E. coli*) bacteria to determine the associated human-health risk of potentially pathogenic (disease-causing) waterborne microorganisms. Coliform bacteria are ubiquitous in the environment and are not always pathogenic. Some strains of *E. coli*, such as O157:H7, can cause severe illness, and the presence of *E. coli* may indicate general fecal contamination because of the strong association of *E. coli* with the intestines of warm-blooded animals. Any detection of *E. coli* in public drinking-water supplies is considered cause for concern and a violation of health standards (U.S. Environmental Protection Agency, 2001), and these EPA criteria can be used to provide a reference point for domestic wells.

Total coliform bacteria were detected in 35 of the 71 groundwater samples (48.6 percent) (Table 1). Total coliform concentrations ranged from less than 1 colony (reported as most probable number [MPN] per 100 milliliters [mL]) to greater than 2,420 MPN/100mL. Of the 35 samples that had total coliform bacteria, 6 samples had detectable *E. coli*, with concentrations ranging from 1 to greater than 2,420 MPN/100mL. The most probable number could not be determined for BR 890, but a present/absent test with the lab determined this sample was absent of total coliform and *E. coli* bacteria, and for the summary Table (Table 1) this sample was represented as less than 1 MPN per 100 milliliters. No value was available for BR 928.

Metals and Trace Elements

Metals and other trace elements typically are present in concentrations less than 1 mg/L in natural waters (Hem, 1985). Most metals and trace elements in groundwater are leached from soil or dissolved from underlying bedrock in minute quantities by groundwater. Some are present in precipitation. Summary statistics for metals and trace elements are listed in Table 1.

The EPA has established MCLs and SMCLs for various metals and trace elements in drinking water (Table 1). Two samples (2.8 percent) exceeded the EPA MCL for arsenic and barium, 8 samples (11.1 percent) exceeded the SMCL for dissolved iron, 22 samples (30.6 percent) exceeded the SMCL for dissolved manganese, and 1 sample exceeded the SMCL for aluminum with a value of 770.0 μg/L. None of the samples exceeded the EPA MCL or SMCL for antimony, beryllium, cadmium, chromium, copper, lead, selenium, silver, or zinc (Table 1).

Arsenic concentrations ranged from less than 0.05 to 23.2 μg/L with a median concentration of 0.63 μg/L (Table 1). Arsenic concentrations were locally elevated and were not consistently related to pH (Figure 4*A*). Two samples, BR 907 and BR 929, had arsenic concentrations exceeding the EPA MCL of 10 μg/L and had a pH of 7.7 and 8.0, respectively. Possible health effects associated with ingestion of drinking water with arsenic in excess of the MCL for many years include skin damage, circulatory system problems, and increased cancer risk. Arsenate ($As^{V}O_4^{3-}$) and arsenite ($As^{III}O_3^{3-}$), which are the predominant forms of arsenic in groundwater, tend to adsorb to a variety of aquifer materials, including iron oxides, aluminum oxides, and clay minerals, at mildly acidic to neutral pH, but not at alkaline pH conditions (Dzombak and Morel, 1990;

Smedley and Kinniburgh, 2002). Furthermore, under strongly acidic or reducing conditions, oxide minerals may become unstable and dissolve, releasing arsenic to solution. More detailed explanation of the effects of pH on sorption processes and the possible relations between redox state and arsenic concentration are presented in a later section of this report.

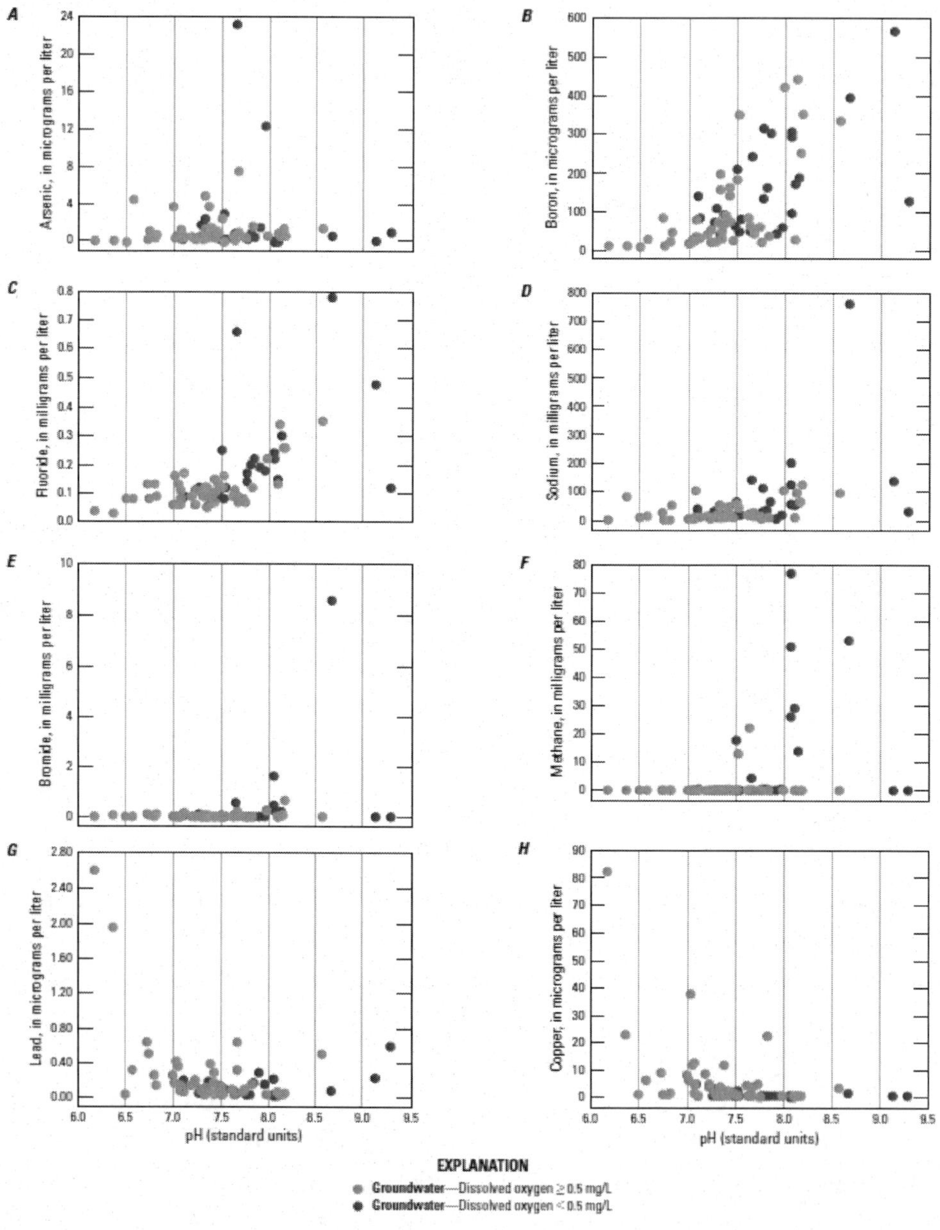

Figure 4. Relation between pH and selected constituents for 72 groundwater samples in Bradford County, Pennsylvania. (≥, greater than or equal to; <, less than).

Although the highest concentrations of bromide (8.60 mg/L) and sodium (760 mg/L) were found in samples having pH greater than 7.5 (Figure 4), most high pH samples did not have elevated concentrations of these constituents. In contrast, the highest concentrations of lead (2.61 µg/L), copper (82.3 µg/L), zinc (981 µg/L), and other trace metals were found in samples that had pH less than 7.5 (Figure 4*G–H*). Generally, decreased concentrations of trace metals with increased pH may be explained by their strong adsorption at alkaline pH by iron oxides, whereas the increased concentrations of trace anions, such as arsenic or boron (present as borate), may result from their desorption at alkaline pH, explained in more detail further on. The increased concentrations of sodium could result from cation-exchange reactions, also explained further on.

Elevated concentrations of iron and manganese in water may impart a bitter taste and stain laundry and plumbing fixtures with a yellowish or brownish-orange color. Concentrations of dissolved iron ranged from less than 4.0 to 1,830 µg/L with a median of 11.34 µg/L. Eight of these samples (11.1 percent) exceeded the EPA SMCL of 300 µg/L for iron. Concentrations of dissolved manganese in water ranged from less than 0.40 to 1,120 µg/L; the median concentration was 10.5 µg/L. Water samples from 22 of the 72 sampled wells (30.6 percent) exceeded the EPA SMCL of 50 µg/L.

Although none of the groundwater samples had concentrations of lead or copper in excess of the respective MCL and SMCL values of 15 and 1,000 µg/L, corrosive water, as previously described by the LI or SICAL less than 0.5, could acquire metals from lead or copper pipes in the household water system. Long-term exposure to excess copper can cause liver and kidney damage, whereas lead can cause neurological problems, especially in young children. Depending on water treatment, the corrosive characteristics may or may not be mitigated. Sampling at the tap would be advisable to evaluate the actual concentrations of constituents in household drinking water and to evaluate the potential effects of water treatment. Lead and copper levels could be much higher if samples were collected before the plumbing system is flushed, as in this study.

Radionuclides

Radionuclides naturally present in rocks and soils may be dissolved or leached into groundwater. Analyses for radioactivity and radionuclides include gross alpha radioactivity, gross beta radioactivity, and dissolved radon-222 (radon gas). Uranium, a radioactive element, also was analyzed in the dissolved form. Naturally occurring radioactivity is described in detail in previous reports on groundwater quality in Wayne, Pike, and Lycoming Counties in Pennsylvania (Senior and others, 2017; Senior and Cravotta, 2017; and Gross and Cravotta, 2017, respectively). Summary statistics for radioactive constituents in groundwater samples from Bradford County as well as the established or proposed EPA MCLs for some of these constituents are given in Table 1.

The activities of radon-222 in water from the 72 sampled wells ranged from 34.4 to 8,540 pCi/L with a median activity of 614 pCi/L (Table 1). The EPA does not regulate radon-222 in drinking water. However, under the framework specified by the 1999 Notice for the Proposed Radon in Drinking Water Rule (Federal Register, 1999), the EPA proposed an alternative maximum contaminant level (AMCL) of 4,000 pCi/L for radon-222 for community water systems that use groundwater for all or some of the supply in states with an enhanced indoor air radon program. For states without an enhanced indoor air program, the EPA proposed an MCL of 300 pCi/L for radon-222. Water samples from 50 of the 72 wells sampled (70.4 percent) exceeded the proposed EPA MCL of 300 pCi/L, and only one sample exceeded the proposed EPA AMCL of 4,000 pCi/L for radon-222.

The gross alpha-particle radioactivity (72-hour count) in water from the 72 sampled wells ranged from −0.33 to 11.2 pCi/L; the median activity level was 1.7 pCi/L (Table 1). Alpha particles are counted at 72 hours and 30 days because some alpha-particle emitters, such as radium-224 (half-life of 3.6 days), would not be present in the 30-day count. Gross alpha-particle activity in the 30-day count was similar to or slightly less than the activity in the 72-hour count. No water sample exceeded the EPA MCL of 15 pCi/L for gross alpha-particle activity (Table 1).

The gross-beta particle radioactivity (72-hour count) ranged from 0.03 to 5.89 pCi/L. Gross beta-particle activity in the 30-day count was similar to or slightly less than the activity in the 72-hour count in about half of the samples (table 1). The concentration of uranium ranged from less than 0.01 to 10.4 μg/L. The median concentration was 0.26 μg/L. No water samples exceeded the EPA MCL of 30 μg/L for uranium.

A subset of eight samples, which had the highest gross-alpha and gross-beta activities of the 72 wells sampled, was analyzed for radium-226 and radium-228. Concentrations (activities) of radium-226 ranged from 0.16 to 2.08 pCi/L in water samples, and activities of radium-228 ranged from −0.02 to 1.54 from the eight wells. None of the samples exceeded the EPA MCL of 5 pCi/L for combined radium-226 and radium-228. The highest activities of radium-226 and radium-228 were measured for the same sample, BR 921. This sample also had the highest 30-day gross alpha and 30-day gross beta activities of the 72 samples reported herein.

Dissolved Methane and Other Naturally Occurring Hydrocarbon Gases

Water sampled from the 72 wells had concentrations of dissolved methane ranging from less than 0.0002 to 77 mg/L, plus lower to nondetectable concentrations of other, more complex hydrocarbon gases, including ethane, ethylene, propane, propylene, isobutane, and N-butane (in order of increasing number of hydrocarbon chains) (Table 1). None of the samples had detectable concentrations of the dissolved hydrocarbons iso-pentane, N-pentane, and hexane.

Detectable levels of methane were found in water from 64 of the 72 wells sampled (88.9 percent) for which concentrations were greater than the reporting level of 0.0002 mg/L (Table 1). Nine of the samples had dissolved methane concentrations that exceeded the Pennsylvania action level of greater than 7 mg/L (Commonwealth of Pennsylvania, 2014). Most samples with detectable concentrations of methane had alkaline pH; the 10 water samples with highest methane concentrations had pH greater than 7.5 (Figure 4*F*). Eight of these 10 samples also had low concentrations of dissolved oxygen (less than 0.5 mg/L) (Figure 4*F*).

Man-Made Organic Compounds

All samples were analyzed for selected man-made organic compounds, including VOCs, glycols and alcohols, and oil and grease (Table 1). Out of these man-made organic compounds, only three were measured in detectable concentrations in 1 of the 72 wells sampled.

VOCs include a wide range of natural and synthetic carbon-based compounds that have high vapor pressure and relatively low solubility in water. VOCs are used in industrial, commercial, and domestic applications and can enter the groundwater as liquid through spills and leaks or by atmospheric deposition. VOCs typically found in groundwater include industrial solvents, fuel hydrocarbons and oxidizers, fumigants, organic synthesis compounds, refrigerants, and disinfection byproducts (trihalomethanes) (Carter and others, 2010). VOCs are the most commonly found contaminant class associated with industrial and commercial sites, dumps, landfills, hazardous waste facilities, and military bases (Zogorski and others, 2006) and are widespread where other human activities occur in developed countries. Historically, many waste chemicals were disposed of indiscriminately, and the widespread use of VOCs has resulted in a considerable mass of material released to the environment. Localized releases of VOCs occur from many sources, including leakage from storage tanks, direct application of pesticides containing VOCs (Barbash and Resek, 1996), septic systems, and leaking sewer systems. VOCs also are released to the atmosphere through engine exhausts, aerosol sprays, leakage of refrigerants, and application of fumigants and pesticides. Trihalomethanes are frequently associated with chlorinated water and chlorinated cleaning products.

The reporting level for VOCs was 0.5 mg/L, which is lower than drinking-water MCLs established for 14 of the 68 analyzed VOCs (U.S. Environmental Protection Agency, 2012a). However, drinking-water MCLs for two compounds— 1,2-dibromo-3-

chloropropane (DBCP) and 1,2-dibromoethane (EDB)—are less than the method reporting level as performed by the laboratory, so for these compounds, a more sensitive method is needed to determine if they are present in concentrations greater than the MCLs (although below the reporting level of 0.5 mg/L). Therefore, with the exception of these two compounds, if VOCs are present in groundwater at concentrations less than the reporting level of 0.5 mg/L, those concentrations would not exceed an established drinking-water standard. MCLs have not been established for VOCs where data are insufficient to evaluate human-health effects.

Of the 68 VOCs analyzed in the 72 groundwater samples, only 3 trihalomethanes were detected above the reporting level: bromodichloromethane (1.04 µg/L), dibromochloromethane (1.58 µg/L), and tribromomethane (0.81 µg/L). All three VOCs were detected in samples from the same well, BR 913, but concentrations of the detected VOCs did not exceed any established EPA drinking-water standard. Although the measured VOC concentrations do not pose any established health risk, the detection of these man-made compounds in groundwater indicates a possible low level of groundwater contamination by human activities.

Glycols and alcohols, and oil and grease, were less than the 5 mg/L reporting level. Although lower concentrations of these constituents may be present in the sampled groundwater, at the time of this study, a more sensitive approved method with lower reporting levels was not available. As of 2018, there are no established EPA drinking-water standards for oil and grease.

Chemical Analysis and Relations among Constituents in Groundwater

Dissolved constituents in groundwater may be derived from atmospheric, geologic, biologic, and anthropogenic sources as the aqueous solution interacts with various materials along transport pathways. Solute concentrations can range widely depending on the presence of constituent elements in the source(s), the extent of contact between water and the source, the aqueous solubility and interactions among the dissolved elements, plus geochemical conditions such as pH and oxidation-reduction (redox) state that affect element form, mobility, and transport in the aqueous environment.

Conceptual Hydrogeochemical Model

A conceptual hydrogeochemical model that indicates spatial relations among the salinity, water types, redox, and methane sources associated with shallow, intermediate, and deep groundwater flow systems for Bradford and surrounding counties is described in Gross and Cravotta (2017). Most wells constructed for domestic use are completed within the local freshwater-flow system. The wells sampled in Bradford County for this study were completed to depths ranging from 42 to 600 feet, with half of the wells drilled to depths of 175 to 400 feet. Several wells with depths ranging from 165 to 300 feet exhibited characteristics of brine-influenced waters with characteristics of elevated SC and TDS dominated by sodium and chloride, plus elevated concentrations of bromide, lithium, and methane, as explained in more detail below.

Relations among pH, Specific Conductance, and Constituent Concentrations

Evaluation of the correlations among chemical constituents and environmental variables provides insight on hydro-chemical processes affecting groundwater chemistry in the study area. The pH, SC, and associated chemical compositions of the 72 groundwater samples collected in Bradford County during 2016 ranged widely, as indicated by pH values from 6.18 to 9.31 and SC from 72.4 to 3,410 µS/cm (Figure 5). Most of the samples had pH values from 7 to 8.2 and SC from 300 to 550 µS/cm. Only 3 of the 72 well-water samples in this study had SC values greater than 1,000 µS/cm. In addition to figure 5, which shows the relations among pH, SC, and multiple inorganic constituents, boxplots are included in the appendix that show the overall relations among individual constituents with pH class interval (Figure 4.1) and SC class interval (Figure 4.2). Additional boxplots show the same constituents as a function of redox class interval (Figure 4.3), bedrock formation (Figure 4.4), and topographic position index (Figure 4.5).

The pH for the 72 groundwater samples was positively correlated with SC and other measures of ionic strength, including TDS (TDScalc and ROE), whereas the pH and SC were negatively correlated with the concentration of DO. With increased pH, the concentrations of several major and trace constituents generally increased, including alkalinity, sodium (Na), lithium (Li), boron (B), fluoride (F), and arsenic (As) (Figure 5, Figure 4.1). Despite positive correlations between SC and the concentrations of chloride (Cl) and bromide (Br), (Figure 5), those constituents were not positively correlated with pH over the entire range of measured values (table 3.2). Concentrations of potassium (K) positively correlated with SC, but were not correlated with pH (table 3.2).

The relation between the pH and concentrations of Ca, Mg, and hardness changes from positive to negative at about pH 7.5 (Figure 5). For pH values less than 7.5, the concentrations of hardness, Ca, Mg, and other cations, including Sr, Na, K, Li, and alkalinity, generally increased with pH. However, for pH values greater than 7.5, the concentrations of hardness, Ca, Mg, Sr, and Ba decreased with increased pH, while SC and concentrations of Na, alkalinity, and TDS continued to increase. Such patterns in pH, SC, and constituent concentrations are consistent with the dissolution of calcite ($CaCO_3 + H^+ = Ca^{2+} + HCO_3^-$) over the range of pH, combined with cation exchange at pH greater than 7.5. Through cation-exchange reactions, the hardness is removed, or softened, as Ca,

Ba, Mg, and Sr ions displace Na ions from exchange sites on clay minerals ($0.5Ca^{2+} + NaX = 0.5CaX_2 + Na^+$) (Appelo and Postma, 2005), as explained in more detail below.

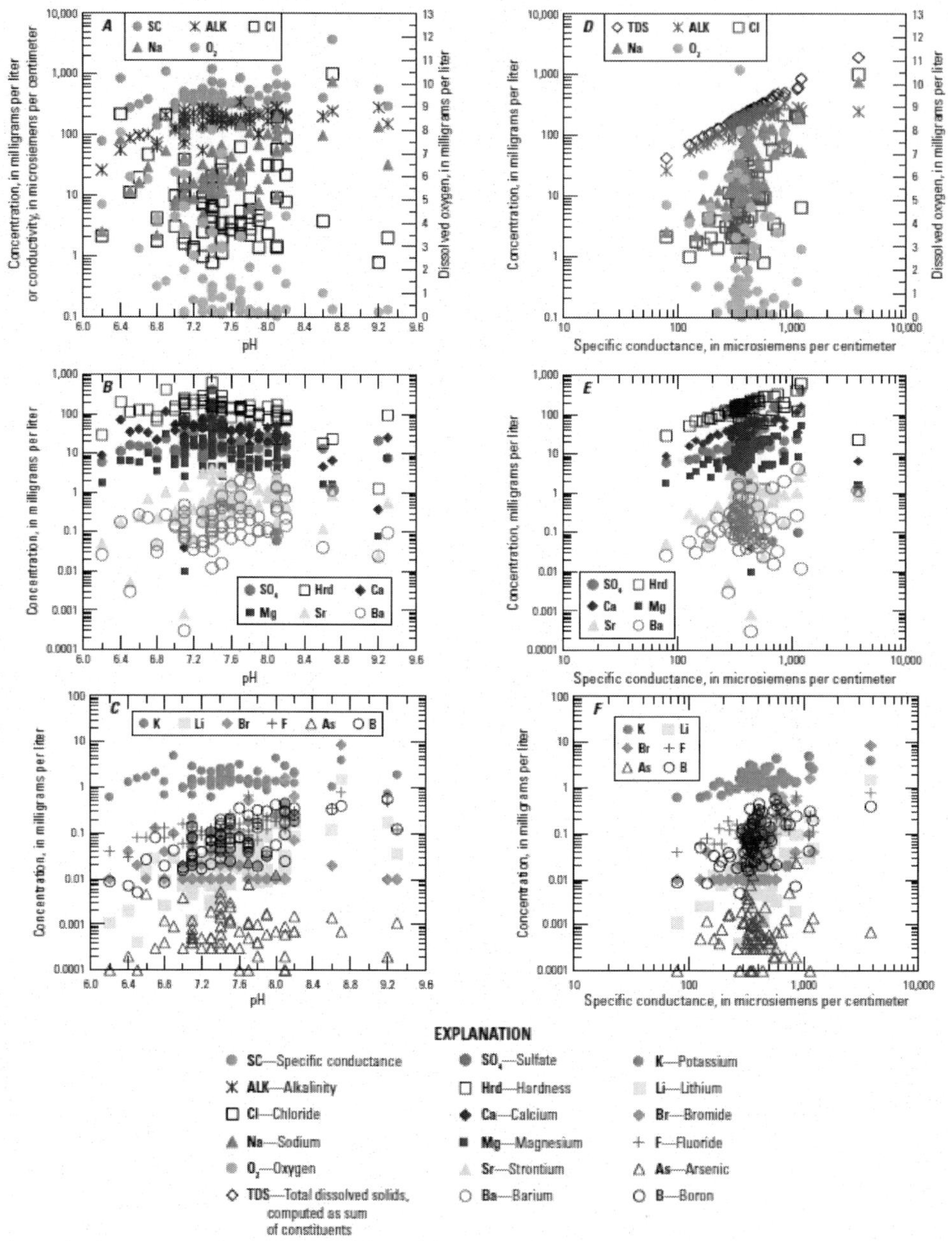

Figure 5. Concentrations of selected constituents in groundwater from Bradford County, Pennsylvania, 2016, compared to: *A–C*, pH; or *D–F*, specific conductance measured in the laboratory. The detection limits for bromide and arsenic were 0.01 and 0.05 micrograms per liter, respectively; symbols plotted at those values were below detection.

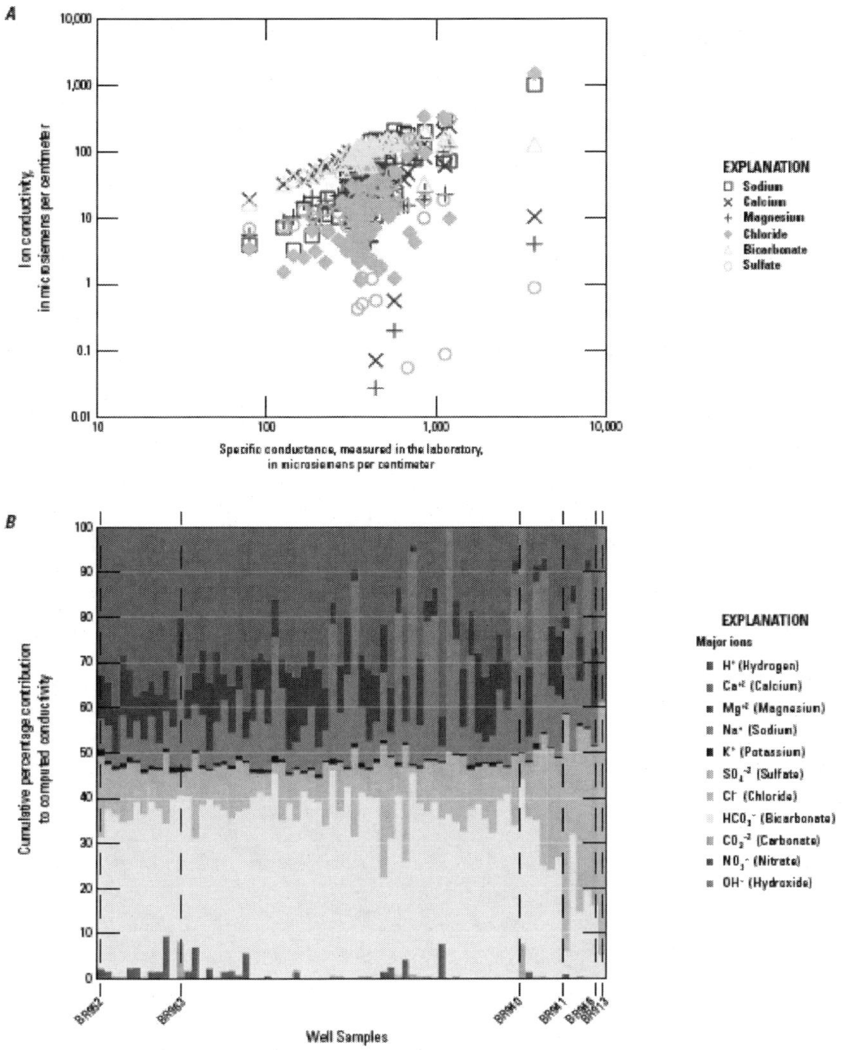

Figure 6. Major ion contributions to specific conductance (SC) for 72 groundwater samples from Bradford County, Pennsylvania, 2016: *A*, comparison of estimated ionic conductivity contributions by sodium, calcium, magnesium, chloride, bicarbonate, and sulfate to measured SC for all 72 samples; and *B*, relative contributions by major ion species to computed SC and selected groundwater samples of representative water types, expressed in percent, in order of increasing SC. Individual ion conductivities estimated from dissolved constituent concentrations as the transport number (the relative contribution of a given ion to the overall conductivity, using the methods of McCleskey and others [2012]) after aqueous speciation calculations with PHREEQC (Parkhurst and Appelo, 2013).

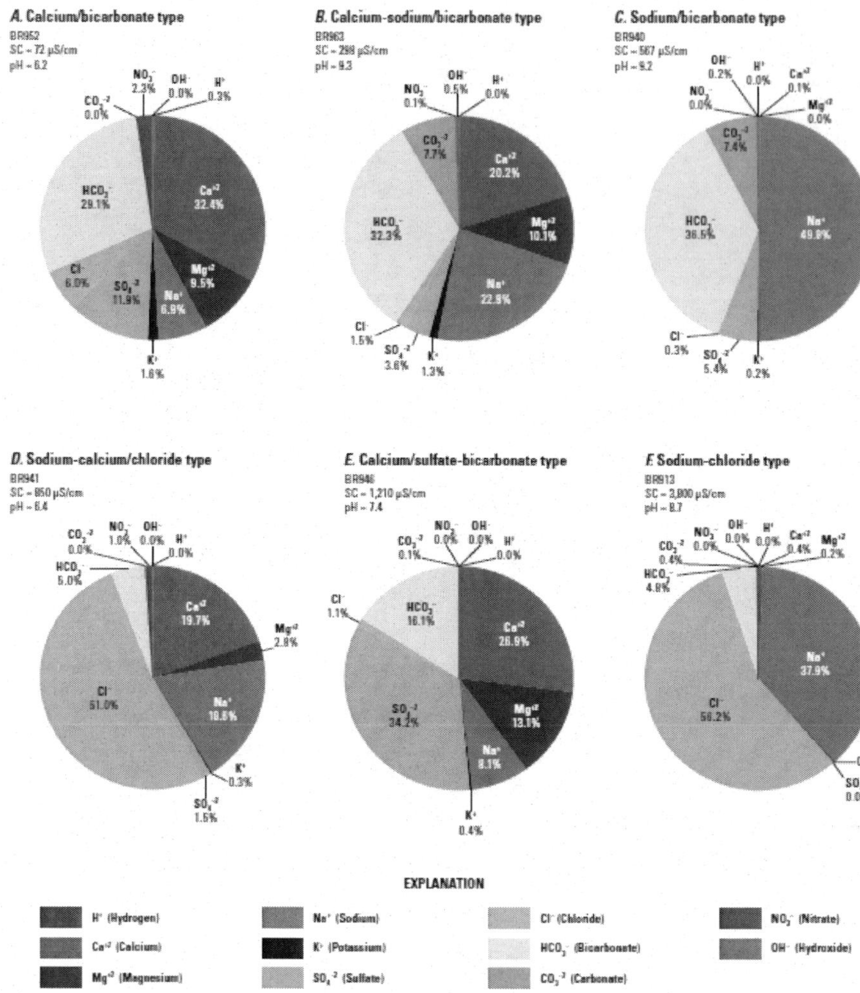

Figure 7. Water types and ionic contributions to specific conductance (SC) computed for selected groundwater samples from Bradford County, Pennsylvania, 2016. Individual ionic conductivities estimated from dissolved constituent concentrations as the transport number (the relative contribution of a given ion to the overall conductivity, using the methods of McCleskey and others [2012]) after aqueous speciation calculations with PHREEQC (Parkhurst and Appelo, 2013). (µS/cm, microsiemens per centimeter).

Boxplots show the constituent concentrations for four different pH class intervals (Figure 4.1). As indicated by the scatter plots that showed constituent concentrations as a function of pH for all 72 groundwater samples (Figure 5), most constituent concentrations increased with pH to values of 7.4 (acidic to neutral range). The most alkaline pH class ($8.0 < pH < 9.4$) had higher median values for SC, TDS, Na, K, alkalinity, Cl, Br, F, B, Li, and methane than the acidic ($6.1 < pH < 6.4$) or near-neutral ($6.5 < pH < 7.4$) pH classes. In contrast, the most alkaline pH class had lower or equal medians for hardness, Ca, Mg, nitrate (NO_3), dissolved oxygen, Zn, U, and Rn-222 compared to the acidic to near-neutral pH classes.

Boxplots also show constituent concentrations for four different SC class intervals (Figure 4.2). As indicated by the scatter plots that showed constituent concentrations as a function of SC for all 72 groundwater samples (Figure 6), most constituent concentrations increased with SC to values approaching 700 µS/cm. The highest SC class (600 < SC < 3,800 µS/cm) had higher median values for pH, SC, TDS, Na, K, alkalinity, Cl, Br, F, B, Li, methane, ammonia (NH_3), and phosphate (PO_4), and lower median values for uranium and Rn-222 than the lower SC classes. Constituents that did not vary with SC and (or) pH may be controlled by other factors, such as redox state.

The constituent concentrations for anoxic, mixed, and oxic redox class intervals also are illustrated as boxplots (Figure 4.3). Although most constituent concentrations did not vary with redox, the medians for several constituents were significantly different between the anoxic and oxic redox classes. The anoxic samples had higher median values for pH, Na, Ba, Sr, Br, F, B, NH_3, Mo, Fe, Mn, and methane, and lower median values for DO, NO_3, U, and Rn-222 than the oxic samples.

Ionic Contributions to Conductivity and Total Dissolved Solids

Despite general correlations between SC and major ion concentrations, the predominance of various ionic contributions to the SC varied widely for samples with comparable values of SC (Figure 6). Calcium and bicarbonate were the predominant ions in most samples, considering the ionic contributions to conductivity or considering the ion concentrations in units of mass, moles, or equivalents. Although sodium and chloride were subordinate in most samples, these constituents were the predominant ions in a few samples, particularly those with SC greater than 1,000 µS/cm, but also in several samples with relatively low values of SC. Sulfate was subordinate to bicarbonate and chloride as a source of conductivity in the samples; its importance generally decreased with increased SC. Contributions by magnesium were comparable to sulfate; however, contributions by nitrate, potassium, hydrogen, hydroxyl, or carbonate ions were relatively minor.

Variations in the major ion species contributions to the laboratory measured SC values for six selected groundwater samples are illustrated in figure 7. To provide context, the samples illustrated in figure 7 are identified among all 72 samples in figure 6. Calcium, bicarbonate, magnesium, sodium, sulfate, and chloride ions are the predominant sources of conductivity for the average of the 72 samples and most other samples having low to moderate SC values (Figure 7*A*, BR 952; Figure 7*B*, BR 963), with minor contributions from nitrate, potassium, and hydrogen ions. Samples with elevated SC values generally have increased contributions from sodium, bicarbonate, and chloride (Figure 7*D*, BR 941; Figure 7*F*, BR 913) and, in some cases, sulfate (Figure 7*E*, BR 946).

Sample BR 952 (Figure 7*A*) is representative of minimally evolved groundwaters with a relatively low SC value of 72 µS/cm. The conductivity of this sample can be produced as the result of rainwater evaporation plus dissolution of various minerals and salts. Sample BR 963 (Figure 7*B*) had higher pH of 9.3 and SC of 298 µS/cm compared to BR 952; this would be classified as a mixed calcium-sodium/bicarbonate water type. In contrast, sample BR 940 (Figure 7*C*), which had pH of 9.2 and SC of 567 μS/cm, had predominant ionic conductivity contributions by sodium and bicarbonate, with lesser contributions by chloride, carbonate, and sulfate ions. The mixed calcium-sodium/bicarbonate and the sodium/bicarbonate waters can be produced as a result of cation exchange processes, where sodium ions in the mineral are exchanged for calcium ions from the water. Chloride is the dominant anion in sample BR 941 (Figure 7*D*), whereas sulfate predominates in BR 946 (Figure 7*K*). These samples have relatively high SC values resulting from the addition of these ions to the groundwater. In the extreme, sample BR 913 (Figure 7*F*), which had pH of 8.7 and the highest measured SC of 3,800 μS/cm, had predominant contributions by sodium and chloride, which indicate a salt source. The predominance of sodium and chloride ionic conductivities was exhibited mainly by those samples with elevated SC but also by a few samples with intermediate SC values, such as sample BR 926, which had pH of 8.2 and SC of 646 μS/cm.

Correlations among Major and Trace Constituents in Groundwater

Constituents that were correlated with one another were "loaded" together and separated from other constituents in a principal component analysis (PCA) model (appendix 3). Six principal components (pH, redox, hardness, chloride-bromide, strontium-barium, molybdenum-arsenic) explain nearly 78.3 percent of the variance in the groundwater dataset and consist of 26 routinely detected constituents (Table 3.1). The association of pH with total dissolved solids, alkalinity, sodium, and other parameters is consistent with the progressive reaction of groundwater with minerals along flow paths, where the older water farther from recharge areas becomes more mineralized, alkaline, and softened. The association of redox-sensitive constituents, including dissolved oxygen, nitrate, iron, manganese, and methane, is interpreted to indicate isolation from the atmosphere, the depletion of oxygen, and the development of reducing conditions that may be attributed to the microbial decomposition of organic matter in the aquifer or introduced to the groundwater. The association of hardness, calcium, magnesium, alkalinity, and sulfate results from the dissolution of calcite, dolomite, gypsum, and, possibly, pyrite or other sulfide minerals. Chloride-bromide associations could indicate the influence of Appalachian Basin brine or other sources of salinity, such as sewage, fertilizer, and (or) road-deicing salt, on the groundwater of associated samples. Lastly, associations of the strontium-barium cations and silica could indicate a common origin or geochemical control, such as silicate or carbonate minerals. Associations of molybdenum-arsenic and lack of correlation with other constituents indicates that variations in the trace-element concentrations are independent of variations in the major ions, SC, pH, or redox.

Relations between Groundwater Quality, Geology, and Topographic Setting

Groundwater acquires solutes through natural and anthropogenic loading of constituents in the recharge area from precipitation, weathering reactions of minerals in the soil and aquifer materials, and constituents applied by human activities at or near the land surface. Additional solutes may be acquired as groundwater flows through the aquifer. Groundwater supplying most domestic wells completed in shallow fractured-bedrock aquifers is derived principally from local recharge and will be influenced by land use and geology near the well.

The bedrock underlying Bradford County mainly consists of clastic sedimentary lithologies that include shale, siltstone, and sandstone. Such clastic rocks are mainly composed of silicate and aluminosilicate minerals, including quartz, feldspar, chlorite, muscovite, and illite, plus minor carbonate, sulfate, sulfide, and oxide minerals that occur as clasts, fracture filling, and cements. Although mineralogy is expected to vary locally, the carbonate, sulfate, and sulfide minerals in the siliciclastic bedrock are prone to weathering in near-surface environments where they will affect pH, hardness, alkalinity, sulfate, and associated solutes. Likewise, the chlorite, muscovite, illite, and various other clay minerals readily accommodate ionic substitutions and are widely recognized to be involved in cation-exchange and sorption processes (Hem, 1985; Appelo and Postma, 2005). The glacial lacustrine and alluvial deposits are generally made up of the same constituents as the local bedrock but also may include some minerals from distant locations.

The pH, SC, and associated constituent concentrations for the Bradford County groundwater samples, classified by the source aquifer bedrock formations, are illustrated as boxplots (Figure 4.4). Of the 72 samples, 47 were obtained from the Lock Haven Formation, 22 from the Catskill Formation, and 3 from the Huntley Mountain Formation. Generally, the medians for pH, SC, TDS, hardness, and most other constituents were significantly lower and the median dissolved oxygen was significantly higher for the waters from the Huntley Mountain Formation compared to those from the Catskill and Lock Haven Formations. Although median values for pH, SC, TDS, hardness, Ca, Cl, and Br were equivalent for the Catskill and Lock Haven Formations, samples from the Lock Haven Formation had slightly higher median values than the Catskill Formation for Mg, Na, alkalinity, SO_4, F, B, NH_3, Fe, Mn, and methane, and lower median values for NO_3, U, and Rn-222.

Major Ion Compositions Indicated by Trilinear Diagrams

Trilinear diagrams indicate the percentage contributions of the major cations and anions, in equivalents (molar concentration multiplied by ionic charge), relative to the total equivalents for cations and anions in a sample. Although similar in concept to the pie graphs showing ionic conductivity contributions to the SC (Figure 7), the factors used to compute the equivalents do not consider ion size and mobility, which are incorporated with the transport numbers for ionic conductivity (giving somewhat different ionic proportions). The corresponding water type is identified on the basis of the predominant (greater than 50 percent) cation(s) and anion(s) shown on the trilinear diagrams (Figure 8). For the Bradford County well-water samples, calcium/bicarbonate type waters were most abundant, with others classified as sodium/bicarbonate or mixed water types, including calcium-sodium/bicarbonate, calcium-sodium/ bicarbonate-chloride, calcium/ sulfate-bicarbonate, sodium/bicarbonate-chloride, sodium/bicarbonatesulfate, or sodium/ chloride types (Figure 8).

Representative samples are identified on the trilinear diagram in figure 8 in order to explain possible origins of the observed water types. Well-water sample BR 952 has near-neutral pH and is classified as calcium/bicarbonate type (Figure 8*A*). A majority of the Bradford County groundwater samples could be classified as calcium/bicarbonate type, which can be produced by the dissolution of calcite ($CaCO_3$) by rainwater (recharge) or groundwater. Sample BR 963 is a mixed calcium-sodium/bicarbonate type, whereas sample BR 940 is a sodium/bicarbonate type. Such water types are somewhat common in Bradford County and likely to form by the dissolution of calcite and (or) dolomite combined with cation exchange; these processes are described in more detail below. Samples plotting away from an apex or within the center part of the diagram are classified as mixed hydro-chemical types. For example, sample BR 946 is classified as a mixed calcium/sulfate-bicarbonate type, BR 941 is classified as a mixed sodium-calcium/chloride type, and sample BR 921 is a mixed calcium-sodium/chloride-bicarbonate type. The mixed water types imply that multiple sources of constituents or processes may be important. Finally, sample BR 913 was a sodium-chloride type, which is unlike any other sample. The predominance of chloride implies a substantial contribution of salt (NaCl, $CaCl_2$) from road-deicing compounds, sewage or animal waste, or possibly brine of geological origin. Predominant water types or hydrochemical facies (Back, 1966) and evolution pathways for mixing of dilute calcium/bicarbonate (Ca/HCO_3) groundwater with road salt, brine, brine combined with cation exchange, or with brine plus calcite dissolution are presented in detail in Gross and Cravotta (2017).

Chloride, Bromide, and Sodium in Groundwater

The concentrations of chloride (0.77–1,020 mg/L), bromide (less than 0.01–8.60 mg/L), and sodium (2.18–760 mg/L) for the 72 groundwater samples collected for this study ranged widely and were positively correlated with one another. In Bradford County, chloride concentrations greater than a few milligrams per liter in shallow groundwater may result from human activities or contributions from naturally occurring deeper, more saline groundwater of regional extent. The elevated chloride concentrations may be associated with elevated concentrations of sodium and, in some cases, nitrate and sulfate, which could indicate influence from human or animal waste. Elevated concentrations of chloride also may be associated with elevated (but two or more orders of magnitude smaller) concentrations of bromide, which could indicate influence from residual brine of geologic origin.

Chloride/bromide ratios can be useful to distinguish different sources of chloride (Davis and others, 1998; Whittemore, 2007; Mullaney and others, 2009). Bromide, like chloride, is a soluble anion that exhibits conservative transport properties and can be used as a tracer, if present at detectable concentrations. Some sources of chloride introduced by human activities into the environment, such as salt (sodium chloride) used for road deicing or present in septic effluent, typically have low amounts of bromide and relatively high chloride/ bromide mass ratios. Recent studies of groundwater quality in nearby Susquehanna County in northeastern Pennsylvania (Warner and others, 2012; Llewellyn, 2014; Siegel and others, 2015) have reported groundwater that has concentrations of chloride and chloride/bromide ratios that indicate possible mixing with higher salinity or brine-type waters; these brine-type waters are postulated to be discharging at the surface from undetermined depths below the freshwater aquifer and mixing with shallow, more dilute groundwater.

The chloride/bromide mass ratios for the 72 Bradford County well-water samples are shown in relation to chloride concentrations in Figure 9, which also shows curves representing compositions resulting from mixing of different proportions of dilute groundwater with (1) relatively pure sodium chloride (NaCl) salt containing only a trace of bromide (such as salt used for road deicing, in water softeners, and present in human and animal waste), or (2) bromide-rich oil and gas well brines (such as those from wells producing gas from the Marcellus Shale). The majority of Bradford County well-water samples, including the samples (BR 941 and BR 957) with the second and third highest chloride concentrations (218 and 215 mg/L, respectively) plot on or near the mixing curve for deicing salt, which may enter groundwater from road runoff or septic effluent. However, several well-water samples (BR 913, BR 921, BR 926, BR 907, and BR 910) had relatively high bromide concentrations (greater than 0.4 mg/L) with low chloride/bromide mass ratios compared to their chloride concentrations of 50 mg/L and higher (Figure 9). These well-water samples plot near or on the mixing curve for oil and gas well brines (Figure 9), suggesting a possible small contribution of chloride from a brine-like source (0.02 percent or less). Also plotting along the brine mixing curve is Salt Spring, a naturally occurring saline spring in Susquehanna County.

Of the 8 water samples that had chloride concentrations greater than 20 mg/L and methane concentrations less than 0.5 mg/L, 2 samples had chloride/bromide ratios that plotted on the road-deicing salt mixing curve, and 4 samples had chloride/bromide ratios that plotted on the brine mixing curve (Figure 9*A–C*). Most of these wells were located in valley or valley sides and had depths ranging from 42 to 250 feet below land surface. Nevertheless, many other wells in the valley settings did not exhibit elevated methane concentrations.

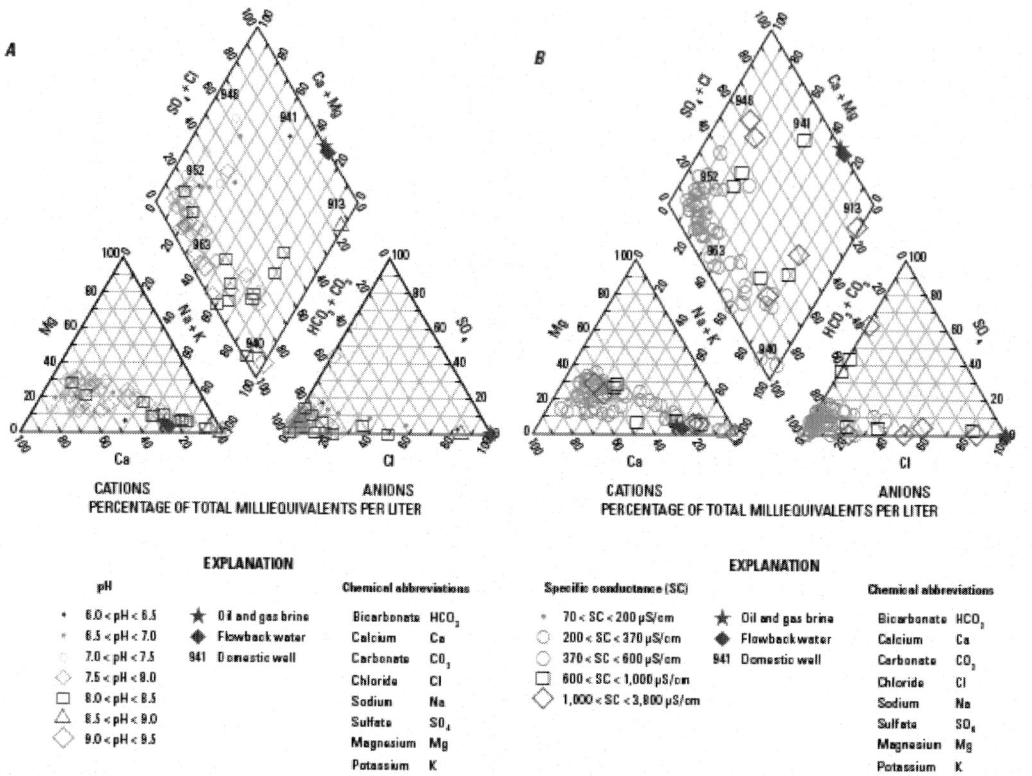

Figure 8. *A*, Predominant water types or hydrochemical facies, and *B*, data for 72 groundwater samples from Bradford County, Pennsylvania, 2016, plus median composition of brine from oil and gas wells in western Pennsylvania (Dresel and Rose, 2010) and flowback water from Marcellus Shale gas wells (Hayes, 2009). Symbols for groundwater scaled by *A*, pH and *B*, specific conductance (SC). (<, less than; µS/cm, microsiemens per centimeter).

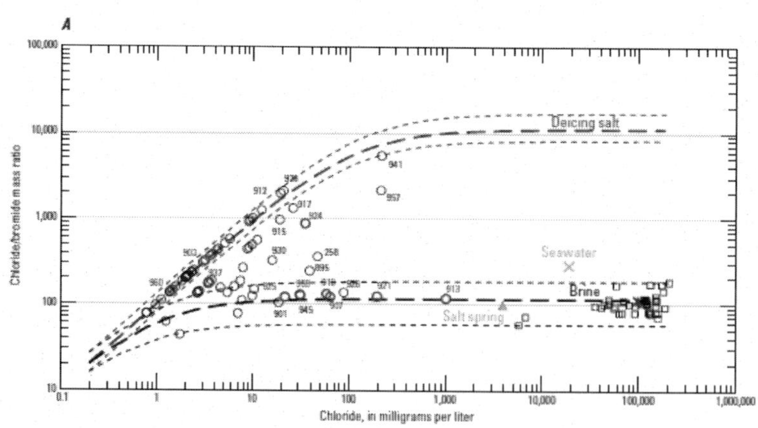

Figure 9. (Continued).

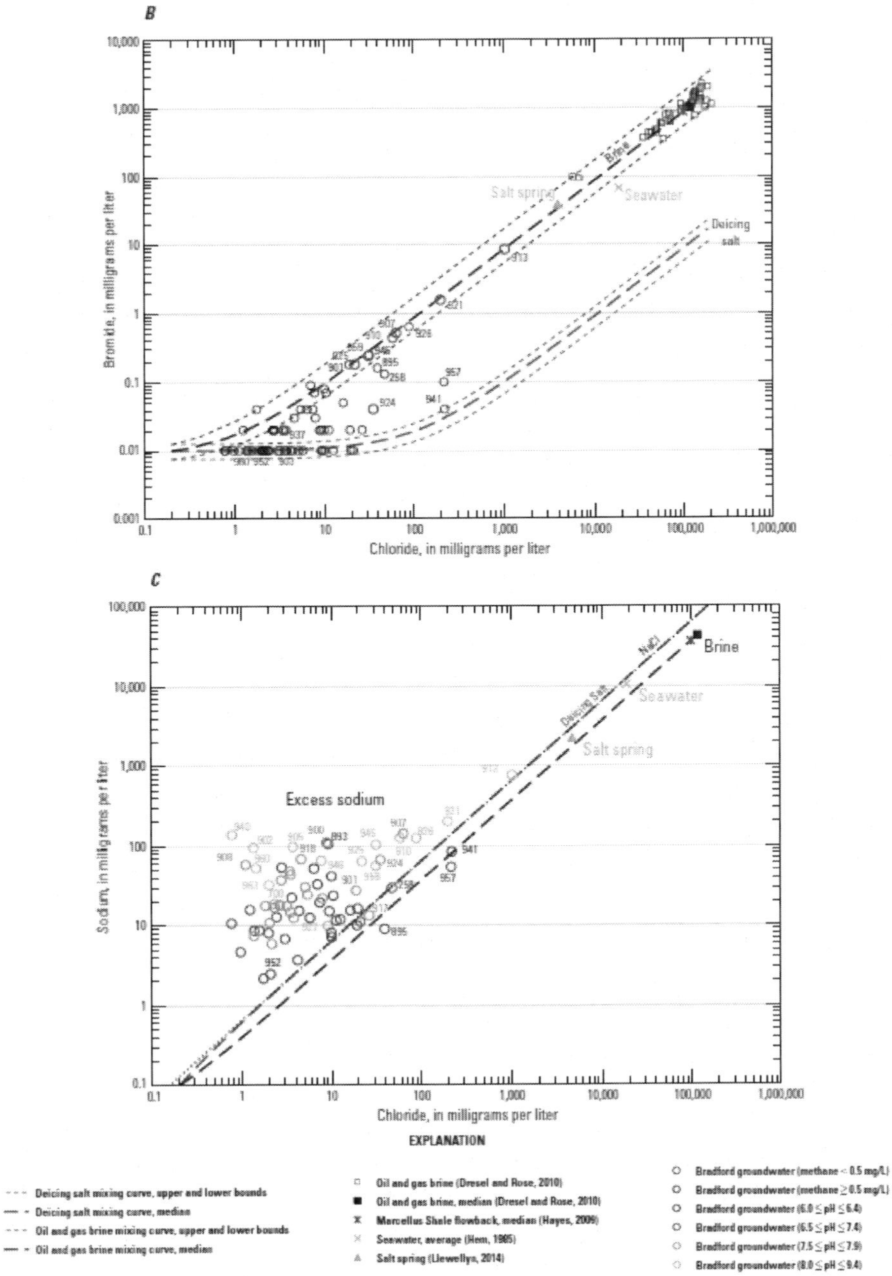

Figure 9. Chloride concentrations compared to *A*, chloride/bromide mass ratios, B, bromide concentrations, and C, sodium concentrations for groundwater from Bradford County, Pennsylvania, 2016, plus median values for Salt Spring (Llewellyn, 2014), flowback waters from Marcellus Shale gas wells (Hayes, 2009), and oil and gas well brines from western Pennsylvania (Dresel and Rose, 2010). Mixing curves computed for initial freshwater with chloride of 0.5 to 0.6 milligrams per liter (mg/L) and bromide of 0.025 to 0.035 mg/L mixed with road deicing salt having a composition of $NaCl_{0.99996}Br_{0.00004}$ (upper curve, after Llewellyn [2014]) or with a median composition of oil and gas well brine (lower curve, after Dresel and Rose [2010]). Selected (numbered) samples have methane concentrations greater than 0.5 mg/L, alkaline pH, or other chemical characteristics that illustrate the variable compositions of the samples from Bradford County. (<, less than; ε, greater than or equal to; δ, less than or equal to).

Many of the Bradford County groundwater samples had chloride/bromide ratios that plot along the mixing curve for road-deicing salt—which is essentially common salt (NaCl)— that is also present in human and animal waste. For example, the water sample from well BR 941, which had the second highest concentration of chloride (218 mg/L=6.15 millimoles per liter (mmol/L)), had a similar molar concentration of sodium (84.6 mg/L=3.68 mmol/L), which could result from the dissolution of pure NaCl by groundwater or recharge water (Figure 9*C*). Nevertheless, compared to the stoichiometry for NaCl, many other samples that had chloride and bromide compositions plotting on the chloride/bromide mixing curve for salt also contained excess sodium (greater than 1:1 [Na]:[Cl] molar ratio) (Figure 9*C*). The excess sodium, derived from mineral sources that do not contain chloride, may result from the release of residual sodium on cation-exchange sites on clay minerals to the groundwater.

The chloride/bromide relations for the Bradford County well-water samples collected for this study generally are similar to those for 20 groundwater samples collected in 2012 in Sullivan County (Sloto, 2013), 95 groundwater samples collected in 2012 and 2015 in Pike County (Senior, 2014; Senior and Cravotta, 2017), and 121 groundwater samples collected in 2013–14 in Wayne County (Senior and others, 2017). Although drilling to access natural gas in the Marcellus Shale produces methane and brine, gas wells have not been drilled to this formation in Pike or Wayne Counties, which are within the Delaware River Basin. The regional, localized occurrence of groundwater throughout northeastern and north-central Pennsylvania that has elevated chloride and chloride/bromide ratios that plot along the mixing curve for brine implies a natural origin for most of the observed chloride, bromide, methane, and associated constituents. However, in rare instances, such as well BR 913, the presence of VOCs and other man-made constituents could indicate an anthropogenic source of the observed methane and brine constituents or connectivity with a deep fracture.

Isotopic Composition, Origin, and Spatial Distribution of Methane

Only the six samples (BR 910, BR 921, BR 903, BR 924, BR 925, and BR 937) with the highest concentrations of methane (13–77 mg/L) were analyzed for the stable isotopic compositions of methane and the associated concentrations of higher-chain hydrocarbon gases (Figure 10). Ethane (C_2H_6) was detected at concentrations ranging from 0.01 to 0.13 mg/L in 4 of 6 samples, but no detectable amount of propane (C_3H_8) or butane (C4H10) was detected in any samples.

The methane in 5 of these 6 groundwater samples (BR 924, BR 903, BR 910, BR 937, and BR 921) had δ13CCH4 values ranging from −38.23 to −65.05 per mil and δDCH4 values ranging from −191.2 to −257.2 per mil, which are consistent with the isotopic compositions reported by Baldasarre and others (2014) for mud-gas logging samples from these geologic units. However, the methane in these groundwater samples has lower (lighter) δ13CCH4 values than those reported by Reese and others (2014) for

Marcellus Shale methane gas samples from Lycoming County. The sixth sample (BR 925) had a methane concentration of 14 mg/L, with an isotopic composition, $\delta 13C_{CH4}$ of −65.1 and δD_{CH4} of −257.2 per mil, and is associated with the higher-chain hydrocarbon ethane that is consistent with methane of microbial origin.

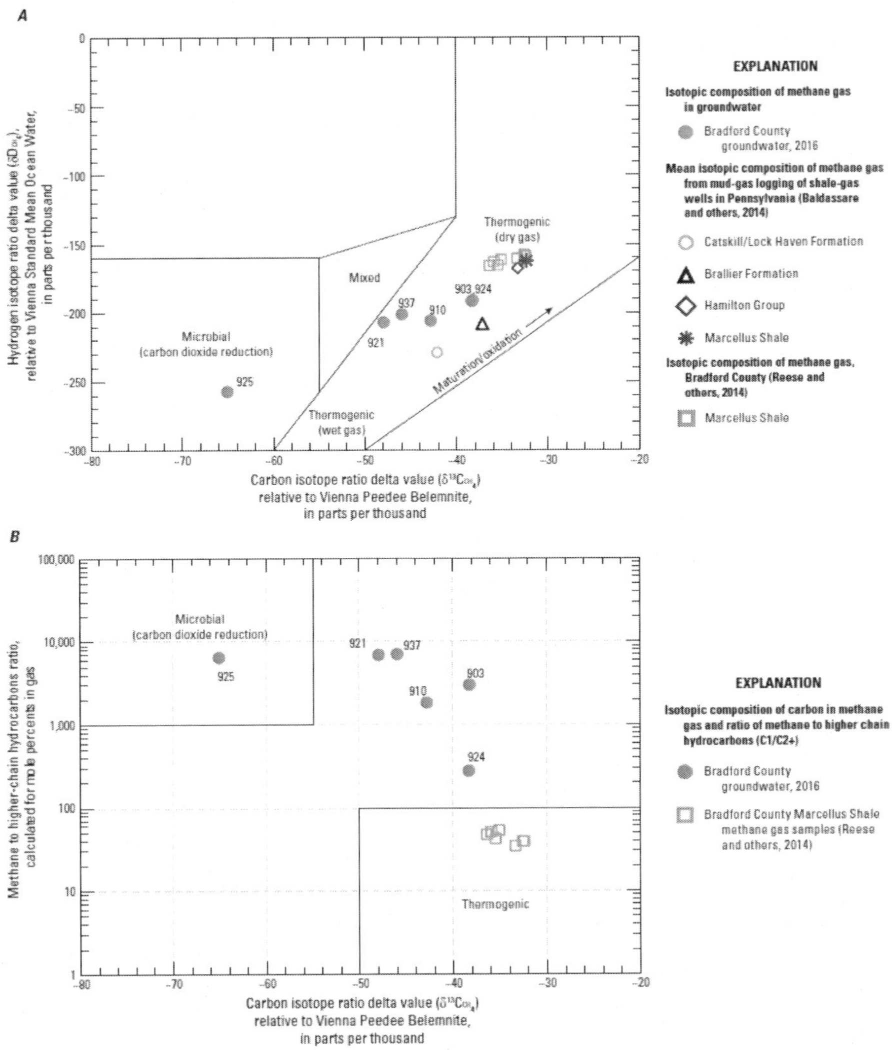

Figure 10. *A*, Isotopic composition of methane in groundwater samples collected from five wells in Bradford County, Pennsylvania, 2016; and *B*, ratio of methane to higher-chain hydrocarbons (C1/C2+) in relation to carbon isotopic composition for methane in these samples. Boundaries for microbial and thermogenic gas types and compositional shift related to gas maturation/oxidation, shown by arrow, from Reese and others (2014, Figure 30, p. 38). Marcellus Shale methane gas data for Bradford County from Reese and others (2014, Table 7) and mud-gas logging data by geologic formation from Baldassare and others (2014, Table 1).

The ratios of methane to ethane plus higher-chain hydrocarbons (C1/C2+) for the other five samples are intermediate between values identified by Reese and others (2014) for microbial or thermogenic methane (Figure 10*B*). These characteristics could indicate a mixed thermogenic and microbial source (carbon-dioxide reduction process) of methane and also could imply that methane of microbial origin has been oxidized (becoming isotopically heavier).Groundwater with relatively elevated methane concentrations (near or above 1 mg/L) also had a chemical composition that differed in some respects (pH, selected major ions, and inorganic trace constituents) from groundwater with low methane concentrations. The five groundwater samples with the highest methane concentrations also had among the highest pH values (greater than 8.2) and elevated concentrations of sodium, lithium, boron, fluoride, arsenic, and bromide. Relatively elevated concentrations of some other constituents, such as barium, strontium, and chloride, commonly were present in, but not limited to, those well-water samples with elevated methane.

Although Llewellyn (2014) described possible associations between elevated methane or brine signatures with deep faulting and associated structural features, the available data for this study indicate that no one physical factor, such as the topographic setting, well depth, or altitude at the bottom of the well, was particularly useful for predicting those well locations with an elevated dissolved concentration of methane. A few parameters stand out as related to topographic position index (TPI) (Figure 4.5); in particular, chloride, bromide, and methane generally increase in the downslope direction. Also, PC4 (Table 3.1, Cl-Br) has progressively higher scores in the downslope direction. These general trends are displayed in the conceptual model that indicates discharge of older, deeper groundwater into the valleys (Gross and Cravotta, 2017).

SUMMARY AND CONCLUSION

In Bradford County, more than half of the residents use groundwater from private domestic-supply wells as their primary drinking source. The quality of private well water is influenced by the regional and local settings, including the surrounding soil, geology, land use, household plumbing, and well construction. The groundwater used for domestic water supply in Bradford County is obtained primarily from shallow bedrock and from unconsolidated (glacial) deposits that overlie bedrock. Historical land use has been predominately forested, agricultural, and residential, but more recently unconventional oil/gas development has also been distributed throughout the various land uses in the county. Despite legislative attempts, Pennsylvania is one of two states in the Nation without statewide water-well construction standards.

To assess the quality of groundwater used for drinking water supplies in Bradford County, samples were collected from 72 domestic wells from May through August 2016 and analyzed for a wide range of constituents that could be evaluated in relation to drinking water health standards, geology, land use, and other environmental factors.

Groundwater samples were analyzed for physical and chemical properties, including major ions, nutrients, bacteria, trace elements, volatile organic compounds, ethylene and propylene glycol, alcohols, gross-alpha/beta-particle activity, uranium, radon-222, and dissolved gases. A subset of samples was analyzed for radium isotopes (radium-226 and -228) and for the isotopic composition of methane.

Results of the 2016 study showed that groundwater quality generally met most drinking-water standards. A subset of samples exceeded primary maximum contaminant levels for total coliform bacteria (49.3 percent), *Escherichia coli* (8.5 percent), and arsenic (2.8 percent); and secondary maximum contaminant levels for sodium (48.6 percent), manganese (30.6 percent), gross alpha and beta activity (16.7 percent), pH (8.3 percent), iron (11.1 percent), total dissolved solids (5.6 percent), chloride (1.4 percent), and aluminum (1.4 percent). Generally, samples that had elevated TDS, chloride, or arsenic concentrations had high pH. Radon-222 activities exceeded the proposed drinking-water standard of 300 pCi/L in 70.4 percent of the samples.

The pH of the groundwater had a large influence on chemical characteristics and ranged from 6.18 to 9.31. Generally, the higher pH samples had higher potential for elevated concentrations of several constituents, including total dissolved solids, sodium, lithium, chloride, fluoride, boron, arsenic, and methane. For the Bradford County well-water samples, calcium/bicarbonate type waters were the most abundant, with others classified as sodium/bicarbonate or mixed water types, including calcium-sodium/bicarbonate, calciumsodium/bicarbonate-chloride, sodium/bicarbonate-chloride, sodium/bicarbonate-sulfate, or sodium/ chloride types. Six principal components (pH, redox, hardness, chloride-bromide, strontium-barium, and molybdenum-arsenic) explain nearly 78.3 percent of the variance in the groundwater dataset.

Groundwater from 12.5 percent of the wells had concentrations of methane greater than the Pennsylvania action level of 7 mg/L; detectable methane concentrations ranged from 0.01 to 77 mg/L. Low levels of ethane (as much as 0.13 mg/L) were present in seven samples with the highest methane concentrations. The isotopic composition of methane in five of these groundwater samples was consistent with the isotopic compositions reported for mud-gas logging samples from these geologic units and a thermogenic source of the methane, and a sixth sample suggests the methane in that sample may be of microbial origin. Well-water samples with the higher methane concentrations also had among the higher pH values and elevated concentrations of sodium, lithium, boron, fluoride, arsenic, and bromide. Relatively elevated concentrations of some other constituents, such as barium and chloride, commonly were present in, but not limited to, those well-water samples with elevated methane.

Four of six groundwater samples with some of the highest methane concentrations had chloride/bromide ratios that indicated mixing with a small amount of brine (0.02 percent or less) similar in composition to those reported for gas and oil well brines in Pennsylvania. In several more eastern Pennsylvania counties where gas drilling is absent, groundwater with comparable chloride/bromide ratios and chloride concentrations have been reported, implying a potential natural source of brine. Most of Bradford County well-water samples have chloride concentrations less than 20 mg/L, and those with higher chloride concentrations have chloride/bromide ratios that indicate anthropogenic sources (such as road-deicing salt and septic effluent) or brine. Brines that are naturally present may originate from deeper parts of the aquifer system, whereas anthropogenic sources are more likely to affect shallow groundwater because they occur on or near the land surface.

The available data for this study indicate that no one physical factor such as the topographic setting, well depth, or altitude at the bottom of the well was particularly useful for predicting those well locations with an elevated dissolved concentration of methane. The 2016 assessment of groundwater quality in Bradford County shows groundwater is generally of good quality, but methane and some constituents that occur in high concentration in naturally occurring brine and produced waters from gas and oil wells may be present at low to moderate concentrations in groundwater in various parts of the aquifer.

Conversion Factors

Multiply	By	To obtain
	Length	
foot (ft)	0.3048	meter (m)
mile (mi)	1.609	kilometer (km)
	Area	
square mile (mi2)	2.590	square kilometer (km2)
	Volume	
gallon (gal)	3.785	liter (L)
	Flow rate	
gallon per minute (gal/min)	0.06309	liter per second (L/s)
	Pressure	
inch of mercury at 60ºF (in Hg)	3.377	kilopascal (kPa)
	Radioactivity	
picocurie per liter (pCi/L)	0.037	becquerel per liter (Bq/L)

Temperature in degrees Celsius (°C) may be converted to degrees Fahrenheit (°F) as follows:

°F = (1.8 ×°C) + 32

Temperature in degrees Fahrenheit (°F) may be converted to degrees Celsius (°C) as follows:

°C = (°F − 32) / 1.8

Datum

Vertical coordinate information is referenced to the North American Vertical Datum of 1988 (NAVD 88).

Horizontal coordinate information is referenced to the North American Datum of 1983 (NAD 83). Altitude, as used in this report, refers to distance above the vertical datum.

Supplemental Information

Specific conductance is given in microsiemens per centimeter at 25 degrees Celsius (µS/cm at 25°C).

Concentrations of chemical constituents in water are given either in milligrams per liter (mg/L) or micrograms per liter (µg/L).

REFERENCES

American Petroleum Institute/America's Natural Gas Alliance, 2013, Bradford and Susquehanna County, Pennsylvania retrospective case study characterization report: Prepared by Battelle for American Petroleum Institute (API) and America's Natural Gas Alliance (ANGA), *Technical Memo* 4, 77 p.

Appelo, C. A. J., and Postma, D., 2005, *Geochemistry, groundwater and pollution*: Leiden, The Netherlands, A. A. Balkema Publishers, 649 p.

Back, W., 1966, *Hydrochemical facies and ground-water flow patterns in northern part of Atlantic Coastal Plain: U.S. Geological Survey Professional Paper 498A*, 42 p., accessed October 2017 at https://doi. org/10.3133/pp498A.

Baldassare, F. J., McCaffrey, M. A., and Harper, J. A., 2014, A geochemical context for stray gas investigations in the northern Appalachian Basin—Implications of analyses of natural gases from Neogene-through Devonian-age strata: *AAPG Bulletin*, v. 98, no. 2, p. 341–372.

Barbash, J. E., and Resek, E. A., 1996, *Pesticides in ground water—Distribution, trends, and governing factors: Chelsea Mich.*, Ann Arbor Press, 598 p.

Belitz, K., Jurgens, B. C., and Johnson, T. D., 2016, Potential corrosivity of untreated groundwater in the United States: U.S. *Geological Survey Scientific Investigations Report* 2016–5092, 16 p., accessed October 2017 at https://doi. org/10.3133/sir20165092.

Botner, E. C., Townsend-Small, A., Nash, D. B., Xu, X., Schimmelmann, A., and Miller, J. H., 2018, Monitoring concentration and isotopic composition of methane in groundwater in the Utica Shale hydraulic fracturing region of Ohio: *Environmental Monitoring and Assessment*, v. 190, no. 6, 15 p.

Boyer, E. W., Swistock, B. R., Clark, J., Madden, M., and Rizzo, D. E., 2012, *The impact of Marcellus gas drilling on rural drinking water supplies: The Center for Rural*

Pennsylvania, 26 p., accessed November 2018 at http://www.rural.palegislature.us/documents/ reports/marcellus_and_drinking_water_2012.pdf.

Brown, M.J., and Margolis, S., 2012, *Lead in drinking water and human blood lead levels in the United States: U.S. Department of Health and Human Services, Centers for Disease Control and Prevention*, accessed November 2018 at https://www.cdc.gov/mmwR/preview/ mmwrhtml/su6104a1. htm.

Carter, J. M., Kingsbury, J. A., Hopple, J. A., and Delzer, G. C., 2010, *Concentration data for anthropogenic organic compounds in groundwater, surface water, and finished water of selected community water systems in the United States*, 2002–10: U.S. Geological Survey Data Series 544, 36 p. [Also available at https://doi.org/ 10.3133/ds 544].

Clune, J. W., and Cravotta, C. A. III, 2018, *Compilation of data not available in the National Water Information System for domestic wells sampled* by the U.S. Geological Survey in Bradford County, Pennsylvania, May–August 2016: U.S. Geological Survey data release, https://doi.org/10.5066/P9VRV6US.

Commonwealth of Pennsylvania, 2014, *The Pennsylvania code chapter 78: Oil and gas wells,* accessed October 2016 at http://www.pacode.com/ secure/data/025/chapter 78/chap78toc.html.

Dai, J., Xia, X., Li, Z., Coleman, D.D., Dias, R.F., Gao, L., Li, J., Deev, A., Li, J., Dessort, D., Duclerc, D., Li, L., Liu, J., Schloemer, S., Zhang, W., Ni, Y., Hu, G., Wang, X., and Tang, Y., 2012, Inter-laboratory calibration of natural gas round robins for δ^2H and $\delta^{13}C$ using off-line and on-line techniques: *Chemical Geology*, v. 310–311, p. 49–55.

Davis, S. N., Whittemore, D. O., and Fabryka-Martin, J., 1998, Uses of chloride/bromide ratios in studies of potable water: *Ground Water*, v. 36, no. 2, p. 338–350.

DeSimone, L. A., 2009, Quality of water from domestic wells in principal aquifers of the United States, *1991–2004: U.S. Geological Survey Scientific Investigations Report 2008– 5227*, 139 p., accessed October 2017 at http://pubs.usgs. gov/sir/2008/5227.

Dresel, P. E., and Rose, A. W., 2010, Chemistry and origin of oil and gas well brines in western Pennsylvania: Pennsylvania Geological Survey, 4th ser, *Open-File Report OFOG 10–01.0*, 48 p.

Durfor, C. N., and Becker, E., 1964, Public water supplies of the 100 largest cities of the United States, 1962: *U.S. Geological Survey Water-Supply Paper 1812*, 364 p. [Also available at https://doi.org/10. 3133/wsp1812].

Dzombak, D. A., and Morel, F. M. M., 1990, *Surface complexation modeling: hydrous ferric oxide*: New York, Wiley Interscience, 393 p.

Federal Register, 1999, *National primary drinking water regulations; radon-222: Federal Register,* v. 64, no. 211, p. 59245–59294, accessed February 2019 at https://www. govinfo.gov/content/pkg/FR-1999-11-02/pdf/99-27741.pdf.

Fishman, M. J., 1993, Methods of analysis by the U.S. Geological Survey National Water Quality Laboratory; determination of inorganic and organic constituents in water and fluvial sediments: U.S. Geological Survey *Open-File Report 93–125*, 217 p. [Also available at https://doi. org/10.3133/ofr93125].

Fishman, M.J., and Friedman, L.C., 1989, Methods for determination of inorganic substances in water and fluvial sediments: *U.S. Geological Survey Techniques of Water-Resources Investigations*, book 5, chap. A1, 545 p. [Also available at https://doi.org/10.3133/twri05A1].

Giddings, T., 2014, Water-well construction standards—Past is prologue?: *Pennsylvania Council of Professional Geologist (PCPG) Newsletter, no. 3, 3 p.*, accessed October 2017 at http://www.pcpg.org/Resources/ Documents/Newsletters/2014/PCPG%20Newsletter%20Q3-2014%201.pdf.

Gross, E.L., and Cravotta, C.A., 2017, Groundwater quality for 75 domestic wells in Lycoming County, Pennsylvania, 2014: U.*S. Geological Survey Scientific Investigations Report 2016–5143*, 74 p., accessed October 2017 at https://doi.org/10.3133/sir20165143.

Hayes, T., 2009, *Sampling and analysis of water streams associated with the development of Marcellus Shale gas: Des Plaines, Ill.*, Report by Gas Technology Institute, for the Marcellus Shale Coalition, accessed October 2017 at http://www.bucknell.edu/script/environmentalcenter/marcellus/default.aspx?articleid=14.

Heilweil, V. M., Grieve, P. L., Hynek, S. A., Brantley, S. L., Solomon, D. K., and Risser, D. W., 2015, Stream measurements locate thermogenic methane fluxes in groundwater discharge in an area of shale-gas development: *Environmental Science & Technology*, v. 49, no. 7, p. 4057–4065.

Heisig, P. M., and Scott, T. M., 2013, Occurrence of methane in groundwater of south-central New York State, 2012— Systematic evaluation of a glaciated region by hydrogeologic setting: *U.S. Geological Survey Scientific Investigations Report 2013–5190*, 32 p., accessed February 2019 at https://doi.org/10.3133/sir20135190.

Helsel, D. R., and Hirsch, R. M., 2002, Statistical methods in water resources: *U.S. Geological Survey Techniques of Water-Resources Investigations*, book 4, chap. A3, 510 p., accessed February 2019 at https://doi.org/10.3133/twri04A3.

Hem, J. D., 1985, Study and interpretation of the chemical characteristics of natural water: *U.S. Geological Survey Water-Supply Paper 2254,* 263 p. [Also available at https:// doi.org/10.3133/wsp2254].

Hirsch, R. M., DeCicco, L., and Lorenz, D., 2015a, dataRetrieval: *Retrieval Functions for USGS and EPA hydrologic and water quality data*, accessed March 13, 2019, at https://rdrr.io/cran/dataRetrieval/.

Hirsch, R. M., DeCicco, L., and Lorenz, D., 2015b, *dataRetrieval reference manual*, accessed March 13, 2019, at https://cran.r-project.org/web/ packages/dataRetrieval/ dataRetrieval.pdf.

Hirsch, R. M., and De Cicco, L. A., 2015, User guide to Exploration and Graphics for RivEr Trends (EGRET) and data Retrieval: R packages for hydrologic data (version 2.0, February 2015): *U.S. Geological Survey Techniques and Methods,* book 4, chap. A10, 93 p. [Also available at https:// doi.org/10.3133/tm4A10].

Hopke, P. K., Borak, T. B., Doull, J., Cleaver, J. E., Eckerman, K. F., Gundersen, L. C. S., Harley, N. H., Hess, C. T., Kinner, N. E., Kopecky, K. J., McKone, T. E., Sextro, R. G., and Simon, S. L., 2000, Health risks due to radon in drinking water: *Environmental Science & Technology*, v. 34, no. 6, p. 921–926.

Jackson, R. B., Vengosh, A., Darrah, T. H., Warner, N. R., Down, A., Poreda, R. J., Osborn, S. G., Zhao, K., and Karr, J. D., 2013, Increased stray gas abundance in a subset of drinking water wells near Marcellus shale gas extraction: *Proceedings of the National Academy of Sciences,* v. 110, no. 28, p. 11250–11255.

Johnson, T., and Belitz, K., 2017, Domestic well locations and populations served in the contiguous U.S.—1990: *Science of the Total Environment*, v. 607–608, p. 658–668.

LaRegina, J., 2013, Testimony of the Pennsylvania Council of Professional Geologists before the *Committee on Environmental Resources and Energy: Pennsylvania House of Representatives Informational Hearing in House Bill 343*, [n.p.], accessed August 31, 2017, at http://wallaby.telicon. com/pa/library/2013/20130417TX.pdf.

Llewellyn, G. T., 2014, Evidence and mechanisms for Appalachian Basin brine migration into shallow aquifers in NE Pennsylvania, USA: *Hydrogeology Journal*, v. 22, no. 5, p. 1055–1066.

Llewellyn, G. T., Dorman, F., Westland, J.L., Yoxtheimer, D., Grieve, P., Sowers, T., Humston-Fulmer, E., and Brantley, S.L., 2015, Evaluating a groundwater supply contamination incident attributed to Marcellus Shale gas development: *Proceedings of the National Academy of Sciences, v. 112*, no. 20, p. 6325–6330.

Lohman, S. W., 1939, Groundwater of north-central Pennsylvania: *Pennsylvania Geological Survey*, no. 6, 219 p.

Low, D. J., and Chichester, D. C., 2006, Ground-water-quality data in Pennsylvania—A compilation of computerized [electronic] databases, 1979–2004: *U.S. Geological Survey Data Series* 150, 22 p., accessed October 2017 at https://doi. org/10.3133/ds150.

McCleskey, R. B., Nordstrom, D. K., Ryan, J. N., and Ball, J. W., 2012, A new method of calculating electrical conductivity with applications to natural waters: *Geochimica et Cosmochimica Acta*, v. 77, p. 369–382.

McMahon, P. B., and Chapelle, F. H., 2008, Redox processes and water quality of selected principal aquifer systems: *Groundwater*, v. 46, no. 2, p. 259–271, accessed

October 16, 2017, at http://onlinelibrary.wiley. com/doi/10.1111/j.1745-6584.2007. 00385.x/abstract.

Miles, C. E., and Whitfield, T. G., comps., 2001, *Bedrock geology of Pennsylvania: Pennsylvania Geological Survey*, 4th ser., dataset, scale 1:250,000.

Molofsky, L. J., Connor, J. A., Farhat, S. K., Wylie, A. S., Jr., and Wagner, T., 2011, Methane in Pennsylvania water wells unrelated to Marcellus shale fracturing: *Oil and Gas Journal*, v. 109, no. 49, accessed October 2017 at https://www.ogj.com/1/vol-109/issue-49/exploration-development/methane-in-pennsylvania-water-full.html.

Molofsky, L. J., Connor, J. A., Wylie, A. S., Wagner, T., and Farhat, S. K., 2013, Evaluation of methane sources in groundwater in northeastern Pennsylvania: *Groundwater*, v. 51, no. 3, p. 333–349.

Mullaney, J. R., Lorenz, D. L., and Arntson, A. D., 2009, Chloride in groundwater and surface water in areas underlain by the glacial aquifer system, northern United States: *U.S. Geological Survey Scientific Investigations Report 2009–5086*, 41 p., accessed February 2019 at https://doi. org/10.3133/sir20095086.

Multi-Resolution Land Characteristics Consortium, 2014, *National Land Cover Database 2011 land cover: Multi-Resolution Land Characteristics Consortium database*, accessed October 15, 2017, at http://www.mrlc.gov.

Murphy, H. M., Thomas, M. K., Schmidt, P. J., Medeiros, D. T., McFadyen, S., and Pintar, K. D. M., 2016, Estimating the burden of acute gastrointestinal illness due to Giardia, Cryptosporidium, Campylobacter, *E. coli* O157 and norovirus associated with private wells and small water systems in Canada: *Epidemiology and Infection*, v. 144, no. 07, p. 1355–1370.

National Agriculture Imagery Program (NAIP), 2016, U.S. Department of Agriculture (USDA) Farm Serve Agency (FSA), Salt Lake City Utah, accessed February 2019 at https://gis.apfo.usda.gov/arcgis/rest/ services/NAIP.

National Research Council, 1998, Volatile organic compounds (VOCs)—Risk assessment of mixtures of potentially carcinogenic chemicals, chap. 6 in *Drinking Water and Health, Selected Issues in Risk Assessment*, volume 9: Washington D.C., National Academies Press, p. 162–167.

National Research Council, 1999, *Arsenic in drinking water*: Washington, D.C., National Academy Press, 273 p.

Osborn, S. G., Vengosh, A., Warner, N. R., and Jackson, R. B., 2011, Methane contamination of drinking water accompanying gas-well drilling and hydraulic fracturing: *Proceedings of the National Academy of Sciences*, v. 108, no. 20, p. 8172–8176.

Otton, J. K., 1992, The geology of radon: *U.S. Geological Survey General Interest Publication*, 29 p., accessed October 2017 at https://doi.org/10.3133/7000018.

Parkhurst, D. L., and Appelo, C. A. J., 2013, Description of input and examples for PHREEQC version 3—A computer program for speciation, batch-reaction, one-dimensional transport, and inverse geochemical calculations: *U.S. Geological Survey Techniques and Methods,* book 6, chap. A43, 497 p., accessed October 2017 at https://doi.org/10.3133/tm6A43.

Patton, C. J., and Kryskalla, J.R., 2011, Colorimetric determination of nitrate plus nitrite in water by enzymatic reduction, automated discrete analyzer methods: *U.S. Geological Survey Techniques and Methods*, book 5, chap. B8, 34 p., accessed October 2017 at https://doi. org/10.3133/tm5B8.

Penn State Extension, 2017, *Iron and manganese in private water systems:* Pennsylvania State University web page, accessed October 23, 2017, at https://extension.psu.edu/iron-and-manganese-in-private-water-systems.

Pennsylvania Bureau of Topographic and Geologic Survey, 1998, Preliminary landform subdivisions of Pennsylvania, 1998: *Pennsylvania Bureau of Topographic and Geologic Survey, Department of Conservation and Natural Resources*, digital data, accessed June 23, 2008, at http://www.pasda. psu.edu.

Pennsylvania Bureau of Topographic and Geologic Survey, 2014, Pennsylvania groundwater information system (PaGWIS) documentation: *Pennsylvania Bureau of Topographicand Geologic Survey,* 4th ser., SQL database, accessed October 2017 from http:// www.dcnr.state.pa.us/topogeo/groundwater/pagwis/index.htm.

Pennsylvania Department of Environmental Protection, 2017, DEP Office of Oil and Gas Management SPUD data: *Pennsylvania Department of Environmental Protection digital database*, accessed May 31, 2018, at http://www.depreportingservices.state. pa.us/ReportServer/Pages/ ReportViewer.aspx?/Oil_Gas/Spud_External_Data.

Pennsylvania General Assembly, 2015a, An act amending Title 27 (Environmental Resources) of the Pennsylvania Consolidated Statutes, in conservation and natural resources, providing for water well construction standards: *Pennsylvania General Assembly House Bill 81,* 8 p., accessed August 31, 2017, at http://www.legis. state.pa.us/cfdocs/billinfo/billinfo.cfm?syear=2015&sind=0&body= H&type=B&bn=0081.

Pennsylvania General Assembly, 2015b, Proposed well construction and decommissioning standards bill: *Pennsylvania General Assembly Senate Bill 653*, 7 p., accessed August 31, 2017, at http://www. legis.state.pa.us/cfdocs/ billinfo/BillInfo.cfm?syear=2015&sind=0&body=S&type=B&bn=653.

Pennsylvania General Assembly, 2016, An act amending Title 27 (Environmental Resources) of the Pennsylvania Consolidated Statutes, in conservation and natural resources, providing for water well construction standards: *Pennsylvania General Assembly House Bill 48,* 8 p., accessed August 31, 2017, at https://www.legis.state.pa.us/cfdocs/billinfo/billinfo.cfm?syear=2015& sind=0&body=H&type=B&bn=0048.

PRISM Group at Oregon State University, 2006a, United States average monthly or annual maximum temperature, 1971–2000: *PRISM Group at Oregon State University digital data,* accessed March 13, 2019, at http://www.prism. oregonstate.edu/.

PRISM Group at Oregon State University, 2006b, United States average monthly or annual minimum temperature, 1971–2000: *PRISM Group at Oregon State University digital data*, accessed March 13, 2019, at http://www.prism. oregonstate.edu/.

Raina, P. S., Pollari, F. L., Teare, G. F., and Goss, M. J., 1999, The relationship between E. coli indicator bacteria in well-water and gastrointestinal illness in rural families: *Canadian Journal of Public Health*, v. 90, no. 3, p. 172.

Reese, S. O., Neboga, V. V., Pelepko, S., Kosmer, W. J., and Beattie, S., 2014, Groundwater and petroleum resources of Sullivan County, *Pennsylvania: Pennsylvania Geological Survey, Water Resource Report 71*, 27 p., accessed February 2019 at http://maps.dcnr.pa.gov/ publications/Default.aspx?id=799.

Reilly, D., Singer, D., Jefferson, A., and Eckstein, Y., 2015, Identification of local groundwater pollution in northeastern Pennsylvania—Marcellus flowback or not?: *Environmental Earth Sciences*, v. 73, no. 12, p. 8097–8109.

Risser, D. W., Thompson, R. E., and Stuckey, M. H., 2008, Regression method for estimating long-term mean annual ground-water recharge rates from base flow in Pennsylvania: *U.S. Geological Survey Scientific Investigations Report 2008–5185*, 23 p., accessed October 2017 at https://pubs. er.usgs.gov/publication/sir20085185.

SAS Institute, Inc., 2012, *SAS 9.4 for Windows*: Cary, N.C., USA, SAS Institute, Inc.

Scott, J. C., 1990, Computerized stratified random site-selection approaches for design of a ground-water-quality sampling network: U.S. *Geological Survey Water-Resources Investigations Report 90–4101,* 109 p., accessed January 1, 1994, at https://doi.org/10.3133/wri 904101.

Senior, L. A., 2009, *Groundwater-quality assessment*, Pike County, Pennsylvania, 2007: U.S. Geological Survey Scientific Investigations Report 2009–5129, 53 p., accessed October 2017 at https://doi.org/10. 3133/sir20095129.

Senior, L. A., 2014, A reconnaissance spatial and temporal baseline assessment of methane and inorganic constituents in groundwater in bedrock aquifers, Pike County, Pennsylvania, 2012–2013: *U.S. Geological Survey Scientific Investigations Report 2014–5117*, 91 p., accessed November 9, 2018, at https://pubs.er.usgs.gov/ publication/ sir20145117.

Senior, L. A., Cravotta, C. A., and Sloto, R. A., 2017, Baseline assessment of groundwater quality in Wayne County, Pennsylvania, 2014: *U.S. Geological Survey Scientific Investigations Report 2016–5073*, 152 p., accessed November 2018 at https://doi.org/10.3133/sir20165073.

Senior, L. A., and Cravotta III, C.A., 2017, Baseline assessment of groundwater quality in Pike County, Pennsylvania, 2015: *U.S. Geological Survey Scientific Investigations*

Report 2017–5110, accessed November 9, 2018, at https://doi.org/10.3133/sir20175110.

Sevon, W. D., and Braun D. D., 1997, Glacial Deposits of Pennsylvania, Map 59: *Pennsylvania Bureau of Topographic and Geologic Survey, Department of Conservation and Natural Resources*, accessed November 11, 2011, at https://www.nrc.gov/docs/ML0707/ ML070720645.pdf.

Sevon, W. D., 2000, Physiographic provinces of Pennsylvania: *Pennsylvania Geological Survey*, 4th ser., Map 13, scale 1:2,000,000.

Siegel, D. I., Smith, B., Perry, E., Bothun, R., and Hollingsworth, M., 2015, Pre-drilling water-quality data of groundwater prior to shale gas drilling in the Appalachian Basin— Analysis of the Chesapeake Energy Corporation dataset: *Applied Geochemistry*, v. 63, no. C, p. 37–57.

Simpson, H., 2004, Promoting the management and protection of private water wells: *Journal of Toxicology and Environmental Health, part A,* v. 67, no. 20–22, p. 1679–1704.

Sloto, R. A., 2013, Baseline groundwater quality from 20 domestic wells in Sullivan County, Pennsylvania, 2012: *U.S. Geological Survey Scientific Investigations Report 2013–5085*, 30 p., accessed October 2017 at https://doi. org/10.3133/sir20135085.

Smedley, P., and Kinniburgh, D., 2002, A review of the source, behaviour and distribution of arsenic in natural waters: *Applied Geochemistry*, v. 17, no. 5, p. 517–568.

Snoeyink, V. L., and Jenkins, D., 1981, *Water chemistry*: New York, John Wiley & Sons, 463 p.

Soeder, D. J., 2017, Unconventional—The development of natural gas from the Marcellus Shale: *The Geological Society of America Special Paper 527*, 143 p.

Swistock, B. R., 2012, Statement of the senior water resources extension associate of the School of Forest Resources, College of Agricultural Sciences, The Pennsylvania State University before the Consumer Affairs Committee: *Pennsylvania House of Representatives Informational Hearing in House Bill 1855*, 4 p., accessed August 31, 2017, at http:// wallaby.telicon.com/pa/library/2012/20120110tq.pdf.

Swistock, B. R., Clemens, S., and Sharpe, W.E., 2009, *Drinking water quality in rural Pennsylvania and the effects of management practices:* The Center for Rural Pennsylvania, 23 p.

Swistock, B. R., and Sharpe, W. E., 2005, The influence of well construction on bacterial contamination of private water wells in Pennsylvania: *Journal of Environmental Health*, v. 68, no. 2, p. 17.

Swistock, B. R., and Sharpe, W. E., 2006, Methane gas and its removal from wells in Pennsylvania: *Pennsylvania State University Cooperative Extension Water Facts* #24, accessed November 2018 at https://extension.psu.edu/methane-gas-and-its-removal-from-water-wells.

Swistock, B. R., Sharpe, W. E., and Robillard, P. D., 1993, A survey of lead, nitrate and radon contamination of private individual water systems in Pennsylvania: *Journal of Environmental Health*, v. 55, no. 5, p. 6.

Taylor, L. E., Werkheiser, W. H., and Kriz, M. L., 1983, Groundwater resources of the West Branch Susquehanna River Basin, Pennsylvania: *Pennsylvania Geological Survey*, 4th ser, 143 p.

U.S. Census Bureau, 2011, *Census 2000 & 2010 redistricting data (Public Law 94-171) summary file*: Prepared by The Pennsylvania State Data Center, Middletown, Pa., accessed October 15, 2017, at http://pasdc. hbg.psu.edu/sdc/pasdc_files/census 2010/032011_release/MCD_PA_PL-2010-2000_Total-Population.xls.

U.S. Environmental Protection Agency, 2001, Total coliform rule—A quick reference guide: *U.S. Environmental Protection Agency, Office of Water*, EPA 816-F-010035, 2 p.

U.S. Environmental Protection Agency, 2003, Drinking water advisory—Consumer acceptability advice and health effects analysis on sodium: *U.S. Environmental Protection Agency EPA 822-R-03-006*, 29 p.

U.S. Environmental Protection Agency, 2012a, Edition of the drinking water standards and health advisories: *U.S. Environmental Protection Agency EPA 822-S-12-001*, accessed October 15, 2017, at http://water. epa.gov/action/advisories/drinking/ upload/dwstandards2012.pdf.

U.S. Environmental Protection Agency, 2012b, Study of the potential impacts of hydraulic fracturing on drinking water resources—Progress report: *U.S. Environmental Protection Agency, EPA/601/R-12/011*, 262 p.

U.S. Geological Survey, variously dated, National field manual for the collection of water-quality data: *U.S. Geological Survey Techniques of Water-Resources Investigations,* book 9, chaps. A1–A7, accessed October 2017 at http:// pubs.water.usgs.gov/twri9A.

U.S. Geological Survey, 2009, 1-Arc second National Elevation Dataset: U.S. Geological Survey, *SDE raster digital data*, accessed October 2017 at https://lta. cr.usgs.gov/NED.

U.S. Geological Survey, 2014, NLCD 2011 land cover (2011 edition): U.*S. Geological Survey, remote-sensing image,* accessed October 2018 at http://www.mrlc.gov.

U.S. Geological Survey, 2018, Water Quality Service: *U.S. Geological Survey Water Services web page,* accessed November 9, 2018, at https://waterservices.usgs.gov/.

Velleman, P. F., and Hoaglin, D., 1981, *Applications, basics, and computing of exploratory data analysis:* Boston, Mass., Duxbury Press, 354 p.

Vidic, R. D., Brantley, S. L., Vandenbossche, J. M., Yoxtheimer, D., and Abad, J. D., 2013, Impact of shale gas development on regional water quality: *Science*, v. 340, no. 6134, 10 p.

Wagner, D. R., 2012, Testimony of the Pennsylvania Council of Professional Geologists chair of Government Affairs Committee before the Consumer Affairs Committee: *Pennsylvania House of Representatives Informational Hearing in House Bill 1855*, [n.p.], accessed August 31, 2017, at http://www.pcpg.org/Resources/ Documents/ PCPG%20HB1855%20Testimony%20with%20Recommendations%20 2012-1-5.pdf.

Warner, K. L., and Ayotte, J. D., 2015, The quality of our Nation's waters—Water quality in the glacial aquifer system, northern United States, 1993–2009: *U.S. Geological Survey Circular 1352*, 128 p., accessed October 2017 at https://doi.org/10.3133/cir1352.

Warner, N. R., Jackson, R. B., Darrah, T. H., Osborn, S. G., Down, A., Zhao, K., White, A., and Vengosh, A., 2012, Geochemical evidence for possible natural migration of Marcellus Formation brine to shallow aquifers in Pennsylvania: *Proceedings of the National Academy of Sciences*, v. 109, no. 30, p. 11961–11966.

Wen, T., Niu, X., Gonzales, M., Zheng, G., Li, Z., and Brantley, S. L., 2018, Big groundwater data sets reveal possible rare contamination amid otherwise improved water quality for some analytes in a region of Marcellus Shale development: *Environmental Science & Technology*, v. 52, no. 12, p. 7149–7159.

Weston Solutions, 2012, Evaluation of geology and water well data associated with the EPA hydraulic fracturing retrospective case study Bradford County, *Pennsylvania: Prepared by Weston Solutions for Chesapeake Energy*, 360 p., accessed October 2017 at https://www.chk.com/news/documents/20120529_chk_westonstudy_ bradfordpa.pdf.

Whittemore, D.O., 2007, Fate and identification of oil-brine contamination in different hydrogeologic settings: *Applied Geochemistry*, v. 22, no. 10, p. 2099–2114.

Williams, J. H., Taylor, L. E., and Low, D. J., 1998, Hydro-geology and groundwater quality of the glaciated valleys of Bradford, Tioga, and Potter Counties, Pennsylvania: Pennsylvania Geological Survey, *Water Resource Report 68,* 89 p.

Yager, R. M., McCoy, K. J., Voss, C. I., Sanford, W. E., and Winston, R. B., 2017, The role of uplift and erosion in the persistence of saline groundwater in the shallow subsurface: *Geophysical Research Letters*, v. 44, no. 8, p. 3672–3681.

Zimmerman, T. M., Zimmerman, M. L., and Lindsey, B. D., 2001, Relation between selected well-construction characteristics and occurrence of bacteria in private household-supply wells, south-central and southeastern Pennsylvania: *U.S. Geological Survey Water-Resources Investigations Report 2001–4206*, 22 p., accessed October 2018 at https:// doi.org/10.3133/wri014206.

Zogorski, J. S., Carter, J. M., Ivahnenko, T., Lapham, W. W., Moran, M. J., Rowe, B. L., Squillace, P. J., and Toccalino, P. L., 2006, Volatile organic compounds in the nation's ground water and drinking-water supply wells: *U.S. Geological Survey Circular 1292*, 101 p., accessed October 2018 at https://pubs.er.usgs. gov/publication/cir1292.

APPENDIXES

Appendix 1

Table 1.1. Compilation of data not available in the National Water Information System, including station name, station identification number, lithologic class, and topographic position index for domestic wells sampled by the U.S. Geological Survey in Bradford County, Pennsylvania, May–August 2016

Station name	Station identification number	Lithologic class	Topographic position index	Station name	Station identification number	Lithologic class	Topographic position index
BR 943	413407076180601	Sandstone	Gentle Slope	BR 897	414717076501801	Sandstone	Valley
BR 956	413515076191101	Sandstone	Lower Slope	BR 954	414747076292001	Mudstone	Gentle Slope
BR 961	413531076323901	Sandstone	Ridges	BR 963	414747076462201	Sandstone	Valley
BR 958	413602076320501	Sandstone	Valley	BR 951	414904076412101	Mudstone	Valley
BR 941	413635076255101	Sandstone	Lower Slope	BR 927	414917076324601	Mudstone	Upper Slope
BR 942	413647076253101	Sandstone	Upper Slope	BR 962	414923076190001	Mudstone	Gentle Slope
BR 924	413658076521001	Mudstone	Lower Slope	BR 894	414930076133001	Sandstone	Gentle Slope
BR 944	413708076131901	Sandstone	Gentle Slope	BR 931	414930076240201	Mudstone	Gentle Slope
BR 932	413743076240801	Sandstone	Upper Slope	BR 935	415029076100101	Mudstone	Valley
BR 959	413828076484101	Sandstone	Valley	BR 935	415029076100101	Mudstone	Valley
BR 919	413834076165001	Sandstone	Ridges	BR 945	415035076493101	Mudstone	Valley
BR 915	413931076120301	Sandstone	Gentle Slope	BR 950	415116076324301	Mudstone	Upper Slope
BR 939	413931076470401	Mudstone	Gentle Slope	BR 929	415118076292401	Mudstone	Gentle Slope
BR 952	413955076305801	Sandstone	Valley	BR 920	415130076185401	Mudstone	Ridges
BR 957	414037076421101	Mudstone	Valley	BR 908	415137076410001	Mudstone	Upper Slope
BR 895	414037076501601	Mudstone	Gentle Slope	BR 904	415141076462001	Sandstone	Gentle Slope
BR 936	414107076185301	Sandstone	Gentle Slope	BR 898	415215076493001	Mudstone	Valley
BR 914	414115076091901	Sandstone	Valley	BR 893	415256076112901	Mudstone	Upper Slope
BR 912	414123076244001	Sandstone	Gentle Slope	BR 925	415353076412201	Mudstone	Valley
BR 923	414138076350601	Sandstone	Gentle Slope	BR 946	415411076175501	Mudstone	Gentle Slope
BR 937	414153076375401	Mudstone	Gentle Slope	BR 948	415414076225901	Mudstone	Lower Slope
BR 960	414158076403901	Mudstone	Upper Slope	BR 949	415430076333901	Mudstone	Steep Slopes
BR 901	414200076264001	Sandstone	Upper Slope	BR 899	415449076513801	Mudstone	Ridges
BR 890	414209076452501	Mudstone	Ridges	BR 947	415508076251901	Mudstone	Upper Slope
BR 896	414234076501301	Sandstone	Valley	BR 922	415608076083801	Mudstone	Valley
BR 906	414319076454401	Mudstone	Gentle Slope	BR 909	415609076294001	Mudstone	Upper Slope
BR 913	414327076140401	Sandstone	Lower Slope	BR 955	415619076475301	Mudstone	Lower Slope
BR 913	414327076140401	Sandstone	Lower Slope	BR 930	415646076202301	Mudstone	Lower Slope
BR 900	414418076313901	Mudstone	Ridges	BR 916	415650076532801	Mudstone	Lower Slope
BR 907	414517076391801	Mudstone	Lower Slope	BR 940	415726076362801	Mudstone	Steep Slopes
BR 911	414522076230801	Mudstone	Lower Slope	BR 926	415729076353901	Mudstone	Valley
BR 258	414542076102801	Sandstone	Lower Slope	BR 921	415807076154401	Mudstone	Valley
BR 892	414552076182201	Mudstone	Upper Slope	BR 918	415819076363401	Mudstone	Lower Slope
BR 902	414608076255701	Mudstone	Upper Slope	BR 917	415921076450601	Mudstone	Ridges
BR 903	414647076224901	Mudstone	Lower Slope	BR 428	415935076184901	Mudstone	Gentle Slope
BR 928	414710076320301	Mudstone	Gentle Slope	BR 953	415937076502901	Mudstone	Gentle Slope

Appendix 2

Table 2.1. Minimum, median, and maximum values of selected characteristics and constituents in groundwater samples collected from 72 domestic wells in Bradford County, Pennsylvania, May–August 2016. Available U.S. Environmental Protection Agency Maximum Contaminant Levels, Action Levels, and Secondary Maximum Contaminant Levels for analyzed constituents (2012 Edition of the Drinking Water Standards and Health Advisories [EPA 822-S-12-001])

Constituent (units)	n	Minimum	Median	Maximum	Results above the reporting level		Results exceeding standard			EPA MCL	EPA SMCL
					Number	Percent	Number	Percent	Graph		
Volatile Organic Compounds											
1,1,1,2-Tetrachloroethane (µg/L)	72	< 0.5	< 0.5	< 0.5	0	0.00	—	—	—	—	—
1,1,1-Trichloroethane (µg/L)	72	< 0.5	< 0.5	< 0.5	0	0.00	0	0		200	—
1,1,2,2-Tetrachloroethane (µg/L)	72	< 0.5	< 0.5	< 0.5	0	0.00	—	—	—	—	—
1,1,2-Trichloroethane (µg/L)	72	< 0.5	< 0.5	< 0.5	0	0.00	0	0		5	—
1,1-Dichloroethane (µg/L)	72	< 0.5	< 0.5	< 0.5	0	0.00	—	—	—	—	—
1,1-Dichloroethene (1,1-Dichloroethylene) (µg/L)	72	< 0.5	< 0.5	< 0.5	0	0.00	0	0		7	—
1,1-Dichloropropene (µg/L)	72	< 0.5	< 0.5	< 0.5	0	0.00	—	—	—	—	—
1,2,3-Trichlorobenzene (µg/L)	72	< 0.5	< 0.5	< 0.5	0	0.00	—	—	—	—	—
1,2,3-Trichloropropane (1,2,3-TCP) (µg/L)	72	< 0.5	< 0.5	< 0.5	0	0.00	—	—	—	—	—
1,2,4-Trichlorobenzene (µg/L)	72	< 0.5	< 0.5	< 0.5	0	0.00	0	0		70	—
1,2,4-Trimethylbenzene (µg/L)	72	< 0.5	< 0.5	< 0.5	0	0.00	—	—	—	—	—
1,2-Dichlorobenzene (o-Dichlorobenzene) (µg/L)	72	< 0.5	< 0.5	< 0.5	0	0.00	0	0		600	—
1,2-Dichloroethane (µg/L)	72	< 0.5	< 0.5	< 0.5	0	0.00	0	0		5	—
1,2-Dichloropropane (µg/L)	72	< 0.5	< 0.5	< 0.5	0	0.00	0	0		5	—
1,3,5-Trimethylbenzene (µg/L)	72	< 0.5	< 0.5	< 0.5	0	0.00	—	—	—	—	—
1,3-Dichlorobenzene (m-Dichlorobenzene) (µg/L)	72	< 0.5	< 0.5	< 0.5	0	0.00	—	—	—	—	—
1,3-Dichloropropane (µg/L)	72	< 0.5	< 0.5	< 0.5	0	0.00	—	—	—	—	—
1,4-Dichlorobenzene (p-Dichlorobenzene) (µg/L)	72	< 0.5	< 0.5	< 0.5	0	0.00	0	0		75	—
2,2-Dichloropropane (µg/L)	72	< 0.5	< 0.5	< 0.5	0	0.00	—	—	—	—	—
2-Butanone (Methyl ethyl ketone, MEK) (µg/L)	72	< 1.0	< 1.0	< 1.0	0	0.00	—	—	—	—	—
2-Chloroethyl vinyl ether (µg/L)	72	< 1.0	< 1.0	< 1.0	0	0.00	—	—	—	—	—
2-Chlorotoluene (µg/L)	72	< 0.5	< 0.5	< 0.5	0	0.00	—	—	—	—	—
4-Chlorotoluene (µg/L)	72	< 0.5	< 0.5	< 0.5	0	0.00	—	—	—	—	—
Acetone (µg/L)	72	< 1.0	< 1.0	< 1.0	0	0.00	—	—	—	—	—
Benzene (µg/L)	72	< 0.5	< 0.5	< 0.5	0	0.00	0	0		5	—
Bromobenzene (µg/L)	72	< 0.5	< 0.5	< 0.5	0	0.00	—	—	—	—	—
Bromochloromethane (µg/L)	72	< 0.5	< 0.5	< 0.5	0	0.00	—	—	—	—	—
Bromomethane (Methyl bromide) (µg/L)	72	< 0.5	< 0.5	< 0.5	0	0.00	—	—	—	—	—
Carbon Disulfide (µg/L)	72	< 0.5	< 0.5	< 0.5	0	0.00	—	—	—	—	—
Chlorobenzene (µg/L)	72	< 0.5	< 0.5	< 0.5	0	0.00	0	0		100	—
CHBrCl2, wu, µg/L (32101)	72	< 0.5	< 0.5	1.04	1	1.39	0	0		—	80
Chloroethane (µg/L)	72	< 0.5	< 0.5	< 0.5	0	0.00	—	—	—	—	—

Constituent (units)	n	Minimum	Median	Maximum	Results above the reporting level		Results exceeding standard			EPA MCL	EPA SMCL
					Number	Percent	Number	Percent	Graph		
Volatile Organic Compounds—Continued											
Chloromethane, wu, µg/L (34418)	72	< 0.5	< 0.5	< 0.5	0	0.00	—	—	—	—	—
cis-1,2-dichloroethene (µg/L)	72	< 0.5	< 0.5	< 0.5	0	0.00	0	0		70	—
cis-1,3-dichloropropene (µg/L)	72	< 0.5	< 0.5	< 0.5	0	0.00	—	—	—	—	—
Dibromochloromethane (µg/L)	72	< 0.5	< 0.5	1.58	1	1.39	0	0		80	—
Dibromomethane (µg/L)	72	< 0.5	< 0.5	< 0.5	0	0.00	—	—	—	—	—
Ethylbenzene (µg/L)	72	< 0.5	< 0.5	< 0.5	0	0.00	0	0		700	—
Hexachlorobutadiene (1,3-Hexachlorobutadiene) (µg/L)	72	< 0.5	< 0.5	< 0.5	0	0.00	—	—	—	—	—
Iodomethane (Methyl iodide) (µg/L)	72	< 0.5	< 0.5	< 0.5	0	0.00	—	—	—	—	—
Isopropylbenzene (Cumene) (µg/L)	72	< 0.5	< 0.5	< 0.5	0	0.00	—	—	—	—	—
m-Xylene plus *p*-xylene (µg/L)	71	< 1.0	< 1.0	< 1.0	0	0.00	0	0		10,000	—
Methyl chloride (Chloromethane) (µg/L)	72	< 0.5	< 0.5	< 0.5	0	0.00	—	—	—	—	—
Methyl *tert*-butyl ether (MTBE) (µg/L)	72	< 0.5	< 0.5	< 0.5	0	0.00	—	—	—	—	—
Naphthalene (µg/L)	72	0.5	< 0.5	< 0.5	0	0.00	—	—	—	—	—
n-Butylbenzene (µg/L)	72	< 0.5	< 0.5	< 0.5	0	0.00	—	—	—	—	—
n-Butyl methyl ketone (µg/L)	72	< 1.0	< 1.0	< 1.0	0	0.00	—	—	—	—	—
n-Propylbenzene (µg/L)	72	< 0.5	< 0.5	< 0.5	0	0.00	—	—	—	—	—
o-Xylene (µg/L)	72	< 0.5	< 0.5	< 0.5	0	0.00	0	0		10,000	—
p-Isopropyltoluene (4-Isopropyltoluene) (µg/L)	72	< 0.5	< 0.5	< 0.5	0	0.00	—	—	—	—	—
sec-Butylbenzene (µg/L)	72	< 0.5	< 0.5	< 0.5	0	0.00	—	—	—	—	—
Styrene (µg/L)	72	< 0.5	< 0.5	< 0.5	0	0.00	0	0		100	—
tert-Butylbenzene (µg/L)	72	< 0.5	< 0.5	< 0.5	0	0.00	—	—	—	—	—
Tetrachloroethene (Perchloroethylene, PCE) (µg/L)	72	< 0.5	< 0.5	< 0.5	0	0.00	0	0		5	—
Tetrachloromethane (µg/L)	72	< 0.5	< 0.5	< 0.5	0	0.00	—	—	—	—	—
Toluene (µg/L)	72	< 0.5	< 0.5	< 0.5	0	0.00	0	0		1,000	—
trans-1,2-Dichloroethene (µg/L)	72	< 0.5	< 0.5	< 0.5	0	0.00	0	0		100	—
trans-1,3-Dichloropropene (µg/L)	72	< 0.5	< 0.5	< 0.5	0	0.00	—	—	—	—	—
Tribromomethane (µg/L)	72	< 0.5	< 0.5	0.81	1	1.39	0	0		60	—
Trichloroethene (TCE, Trichloroethylene) (µg/L)	72	< 0.5	< 0.5	< 0.5	0	0.00	0	0		5	—
Trichloromethane (µg/L)	72	< 0.5	< 0.5	< 0.5	0	0.00	0	0		80	—
Trihalomethanes (THMs), summation (µg/L)	72	< 2.0	< 2.0	3.43	1	1.39	—	—	—	—	—
Vinyl Acetate (µg/L)	72	< 0.5	< 0.5	< 0.5	0	0.00	—	—	—	—	—
Vinyl Chloride (Chloroethene) (µg/L)	72	< 0.5	< 0.5	< 0.5	0	0.00	0	0		2	—
Xylene (µg/L)	72	< 1.5	< 1.5	< 1.5	0	0.00	—	—	—	—	—

EXPLANATION

(shaded) **No exceedance**

[n, number of results; EPA, U.S. Environmental Protection Agency; MCL, Maximum Contaminant Level; SMCL, Secondary Maximum Contaminant Level; µg/L, micrograms per liter; <, less than; —, no MCL, Action Level, or SMCL established].

Appendix 3

Principal component analysis (PCA) provides insight on hydrochemical processes affecting groundwater chemistry in the study area by indicating intercorrelations among chemical constituents and environmental variables. Six principal components (PCs) explain nearly 78.3 percent of the variance in the groundwater dataset and consist of 26 routinely detected constituent loadings (Table 3.1). As explained by Thyne and others (2004), the loading value is comparable to the correlation coefficient between a constituent and the PC, whereas the communality and eigenvalues indicate the degree to which the PCA explains the total variability of the elements considered in the analysis. Associations of additional chemical and physical variables excluded from the model are indicated by the Spearman-rank coefficient of correlation of these variables with the PC scores (Table 3.2); significant correlations are listed below the six PCs included in Table 3.1, which are identified as principal component 1 (PC1) through principal component 6 (PC6).

PC1 (pH), which explains 36.1 percent of the variance in the data, has positive loadings by boron, lithium, sodium, alkalinity, pH (field), fluoride, specific conductance (SC) (laboratory), ammonia, and methane, with negative loading by radon-222, nitrate, calcium, and uranium (Table 3.2). Scores on PC1 are positively correlated with total dissolved solids (TDS; Table 3.2). The negative associations of PC1 with calcium and radon 222 and positive associations with pH, alkalinity, lithium, and sodium are consistent with the progressive reaction of groundwater with minerals along flow paths, where the older water further from recharge areas becomes more mineralized, alkaline, and "softened." Samples classified as anoxic generally had higher, positive scores on PC1 compared to those classified as mixed or oxic, plus PC1 scores were generally greatest for samples from the Lock Haven formation and least for Huntley Mountain, as shown by Appendix Figures 4.3 and 4.4. High positive scores on PC1 generally may be attributed to the progressive weathering of calcite and dolomite (carbonate minerals) combined with cation-exchange processes. The naturally occurring cation-exchange reactions liberate sodium and other alkali earth cations, such as lithium, while removing calcium and magnesium (hardness), much like a water-softening treatment system (Hem, 1985; Poth, 1962). The removal of calcium and magnesium from solution leads to undersaturation of the groundwater with respect to calcite and dolomite, thus promoting additional dissolution of the carbonate minerals and progressive increases in pH and alkalinity along the flow path. The resultant sodium-bicarbonate waters tend to have alkaline pH (>8) values and high positive scores on PC1 (Figure 3.2). Scores on PC1 also increased with SC class interval (Figure 4.2).

PC2 (Redox), which explains 13.5 percent of the variance in the data, has positive loadings by manganese, iron, ammonia, and methane, and negative loadings by dissolved oxygen, nitrate, and radon-222 (Table 3.2). Scores on PC2 are negatively correlated with selenium, copper, and redox potential (Eh). High scores on PC2 are interpreted to indicate isolation from the atmosphere, the depletion of oxygen, and the development of reducing conditions that may be attributed to the microbial decomposition of organic matter in the aquifer or introduced to the groundwater. Samples classified as having anoxic or mixed redox characteristics generally had positive scores on PC2, whereas samples classified as oxic generally had negative scores (Figure 4.3). The negative correlations of copper and selenium with PC2 indicate decreased mobility of these constituents where concentrations of dissolved iron and manganese are elevated, possibly because of adsorption by hydrous ferric oxide (HFO) minerals, including goethite ($FeOOH$) or ferrihydrite ($Fe(OH)_3$), that could be present along with the dissolved iron.

PC3 (Hardness), which explains 8.7 percent of the variance in the data, has positive loadings by sulfate, magnesium, calcium, SC (laboratory), silica, and alkalinity. Scores on PC3 are positively correlated with hardness, nickel, and TDS. High scores on PC3 can be attributed to the dissolution of calcite, dolomite, gypsum, and, possibly, pyrite or other sulfide minerals, without the cation-exchange softening effects (alkaline pH) indicated by high scores on PC1. Median PC3 scores decreased in the order of Lock Haven, Catskill, and Huntley Mountain Formations, as shown by Appendix Figure 4.4.

PC4 (Chloride-Bromide), which explains 7.2 percent of the variance in the data, has positive loadings by chloride, bromide, SC (laboratory), and sodium and is positively correlated with TDS (Table 3.2). The positive association between chloride, bromide, and sodium could indicate the influence of Appalachian Basin brine or other sources of salinity, such as sewage, fertilizer, and (or) road-deicing salt, on the groundwater of associated samples. Scores on PC4 increased progressively with SC class interval (Figure 4.2) and generally were greater for samples from the Catskill Formation than those from the Lock Haven or Huntley Mountain bedrock formations (Figure 4.4).

PC5 (Strontium-Barium), which explains 6.7 percent of the variance in the data, has positive loadings by strontium, potassium, barium, silica, and lithium (Table 3.2) and was positively correlated with gross beta activity. The positive associations of the cations and silica could indicate a common origin or geochemical control, such as silicate or carbonate minerals. For example, potassium, barium, and strontium commonly substitute for calcium in carbonate (aragonite) and for one another in sulfate (barite-celestine) minerals, which could be possible sources or sinks of the cations (Hanor, 1968; Hanshaw and Back, 1979). Additionally, these cations may be retained by clay minerals involved in cation exchange (Appelo and Postma, 2005) and, thus, may be released as the exchange sites become depleted in sodium and enriched in calcium and magnesium.

PC6 (Molybdenum-Arsenic), which explains 6.0 percent of the variance in the data, has positive loadings by molybdenum, arsenic, uranium, and radon-222. The positive associations of these trace elements and lack of correlation with other constituents, indicates that variations in the trace-element concentrations are independent of variations in the major ions, SC, pH, or redox.

References

Appelo, C. A. J., and Postma, D., 2005, *Geochemistry, groundwater and pollution:* Leiden, The Netherlands, A. A. Balkema Publishers, 649 p.

Hanor, J. S., 1968, Frequency distribution of compositions in the barite-celestite series: *American Mineralogist*, v. 53, p. 1215–1222.

Hanshaw, B. B., and Back, W., 1979, Major geochemical processes in the evolution of carbonate—Aquifer systems: Contemporary Hydrogeology—*The George Burke Maxey Memorial* Volume, v. 43, no. 1, p. 287–312.

Hem, J. D., 1985, Study and interpretation of the chemical characteristics of natural water: *U.S. Geological Survey Water-Supply Paper* 2254, 263 p., [Also available at https:// doi.org/10.3133/wsp2254].

Poth, C. W., 1962, The occurrence of brine in western Pennsylvania: Pennsylvania Geological Survey 4th ser., *Mineral Resources Report* 47, 53 p.

SAS Institute, Inc., 2012, SAS 9.4 for Windows: Cary, N.C., USA, SAS Institute, Inc.

Thyne, G., Guler, C., and Poeter, E., 2004, Sequential analysis of hydrochemical data for watershed characterization: *Ground Water*, v. 42, no. 5, p. 711–723.

Table 3.1. Principal components analysis model of major factors controlling the chemistry of groundwater and statistically significant Spearman rank correlations of other variables with those factors determined from chemical properties of and dissolved concentrations of constituents in groundwater samples from 72 wells, Bradford County, Pennsylvania, 2016

Principal component	PC1 (pH)	PC2 (Redox)	PC3 (Hardness)	PC4 (Chloride-Bromide)	PC5 (Strontium-Barium)	PC6 (Molybdenum-Arsenic)	Communality
Constituent loadings							
Boron (B)	87 *	26 +	−7	9	26 +	1	0.912
Lithium (Li)	84 *	9	−13	16	36 *	3	0.891
Sodium (Na)	79 *	25 +	−4	39 *	10	2	0.855
Alkalinity (ALK)	70 *	31 +	41 *	21	3	5	0.805
pH, Field (pHF)	69 *	30 +	−32 +	−13	10	20	0.735
Fluoride (F)	68 *	22	−35 +	11	3	−10	0.660

Principal component	PC1 (pH)	PC2 (Redox)	PC3 (Hardness)	PC4 (Chloride-Bromide)	PC5 (Strontium-Barium)	PC6 (Molybdenum-Arsenic)	Communality
Specific conductance, lab (SCL)	56 *	21	50 *	53 *	0	−11	0.907
Radon 222 (Rn222)	−68 *	−37 *	−11	4	3	37 *	0.756
Manganese (Mn)	6	85 *	31	17	8	3	0.863
Iron (Fe)	24	76 *	4	22	−3	−17	0.710
Ammonia (NH3N)	55 *	68 *	0	20	19	−15	0.868
Methane (Methane)	41 *	67 *	−16	26	24	−9	0.781
Nitrate (NO3N)	−46 *	−66 *	2	8	−16	5	0.681
Dissolved oxygen (DOX)	−22	−74 *	13	1	−16	−18	0.679
Sulfate (SO4)	7	−9	84 *	−2	−22	7	0.779
Magnesium (Mg)	−18	9	80 *	1	28	2	0.767
Calcium (Ca)	−42 *	5	73 *	20	22	21	0.853
Chloride (Cl)	0	14	3	89 *	10	−6	0.833
Bromide (Br)	31 +	14	−1	81 *	10	−1	0.779
Strontium (Sr)	29 +	32 +	1	6	77 *	10	0.793
Potassium (K)	31 +	−3	20	18	72 *	−4	0.696
Barium (Ba)	3	19	−46 *	35 +	62 *	3	0.755
Silica (SiO2)	−1	20	48 *	−26	57 *	2	0.657
Molybdenum (Mo)	−2	15	4	−4	4	93 *	0.886
Arsenic (As)	8	−9	8	−4	6	78 *	0.628
Uranium (U)	−53 *	−31 +	11	−2	−15	65 *	0.832
Eigenvalue							
Eigenvalue	9.40	3.52	2.27	1.88	1.74	1.56	-
Percent Variance Explained	36.1	13.5	8.7	7.2	6.7	6.0	-
Cumulative Percent Variance Explained	36.1	49.7	58.4	65.7	72.3	78.3	-
Significant Spearman correlations ($p < 0.001$)							
Total dissolved solids (TDS)	57	-	46	48	-	-	-
Phosphate (PO4P)	38	-	-	-	-	-	-
Ethane (Ethane)	37	43	−33	-	-	-	-
Selenium (Se)	-	−48	-	36	-	-	-
Copper (Cu)	−44	−57	-	-	-	-	-
Significant Spearman correlations ($p < 0.001$)—Continued							
Redox potential (EHmv)	-	−64	-	-	-	-	-
Hardness (HRDcalc)	-	-	79	-	-	-	-
Nickel (Ni)	-	-	48	-	-	-	-
Gross beta radioactivity, 30-day count, water, filtered, Cesium-137 curve, picocuries per liter (BetaCs30)	-	-	-	-	51	-	-
Gross beta radioactivity, 72-hour count, water, filtered, Cesium-137 curve, picocuries per liter (BetaCs72)	-	30	-	-	47	-	-

Table 3.1. (Continued)

Principal component	PC1 (pH)	PC2 (Redox)	PC3 (Hardness)	PC4 (Chloride-Bromide)	PC5 (Strontium-Barium)	PC6 (Molybdenum-Arsenic)	Communality
Gross alpha radioactivity, 30-day count, water, filtered, Th-230 curve, picocuries per liter (AlphaTh30)	-	-	-	-	38	-	-
Gross alpha radioactivity, 72-hour count, water, filtered, Th-230 curve, picocuries per liter (AlphaTh72)	-	-	-	-	-	34	-
Cadmium (Cd)	−32	-	-	-	-	-	-
Lead (Pb)	−32	−31	-	-	-	-	-
Antimony (Sb)	-	−31	-	-	-	-	-
Well depth (WELLZ)	31	−32	-	-	-	-	-
pH, Lab (pHL)	69	-	−33	-	-	-	-
Residue on evaporation 180C (ROE180)	58	-	43	48	-	-	-
Specific conductance, field (SCF)	53	-	47	53	-	-	-
Aluminum (Al)	-	-	-	-	-	-	-
Beryllium (Be)	-	-	-	-	-	-	-
Cobalt (Co)	-	-	-	-	-	-	-
Land-surface altitude (LSALT)	-	-	-	-	-	-	-
Propane (Propane)	-	-	-	-	-	-	-
Total coliform (Tcoli)	-	-	-	-	-	-	-
Temperature (TEMPC)	-	-	-	-	-	-	-
Depth to water (WATRZ)	-	-	-	-	-	-	-
Well bottom altitude (WELLZALT)	-	-	-	-	-	-	-
Zinc (Zn)	-	-	-	-	-	-	-

[PC1–PC6, principal components 1 through 6 with major factor indicated in parentheses; Varimax rotation pattern for rank-transformed data (SAS, 2012); minimum eigenvalue > 1; loading values for constituents included in model and Spearman correlation coefficients for constituents excluded from model multiplied by 100 and rounded; *, indicates highly significant loading or correlation ($p < 0.001$); + indicates significant loading or correlation ($p < 0.01$); -, indicates insignificant correlation or not applicable].

Table 3.2. Spearman rank correlation coefficient (r) matrix for groundwater chemical data Bradford County, Pennsylvania, 2016

Parameter name	Parameter abbreviation	PC1	PC2	PC3	PC4	PC5	PC6	TEMPC	DOX	EHmv	LSALT	WELLZALT	WELLZ	WATRZ	pH	pHL	SCL	SCF	ROE180	TDS	HRDcalc	Ca	Mg	K	Na	ALK	Br	Cl	F	SiO_2	SO_4	NH_3N	NO_3N	PO_4P	Tcoli	Al	Ba	Be	Cd	Co	Cr	Fe	Pb	Li	Mn	Mo	Ni	Sr	Zn	Sb	As	B	Se	AlphaTh30	AlphaTh72	BetaCs30	BetaCs72	Rn222	U	Methane	Ethane	Propane
Principal component 1	PC1	100	-	-	-	-	-	-	-	-	-	-	31	-	69	69	55	53	58	57	-	-35	-	40	79	72	32	-	66	-	-	55	-48	38	-	-	-	-	-32	-	-44	-	-32	85	-	-	-	39	-	-	-	89	-	-	-	-	-	-66	-52	43	37	-
Principal component 2	PC2	-	100	-	-	-	-	-	-75	-64	-	-	-33	-	30	-	-	-	-	-	-	-	-	-	30	33	-	-	-	-	-	70	-66	-	-	-	-	-	-	-	-57	76	-31	-	87	-	-	33	-	-33	-	30	-48	-	-	-	30	-38	-34	71	43	-
Principal component 3	PC3	-	-	100	-	-	-	-	-	-	-	-	-	-	-33	-33	46	47	43	46	79	72	80	-	-	36	-	-	-	45	85	-	-	-	-	-	-44	-	-	-	-	-	-	-	-	-	48	-	-	-	-	-	-	-	-	-	-	-	-	-	-33	-
Principal component 4	PC4	-	-	-	100	-	-	-	-	-	-	-	-	-	-	-	53	53	48	48	-	-	-	-	37	-	83	89	-	-	-	-	-	-	-	-	31	-	-	-	-	-	-	-	-	-	-	-	-	-	-	-	36	-	-	-	-	-	-	-	-	-
Principal component 5	PC5	-	-	-	-	100	-	-	-	-	-	-	-	-	-	-	-	-	-	-	-	-	-	71	-	-	-	-	-	59	-	-	-	-	-	-	58	-	-	-	-	-	-	42	-	-	-	77	-	-	-	33	-	38	-	51	47	-	-	-	-	-
Principal component 6	PC6	-	-	-	-	-	100	-	-	-	-	-	-	-	-	-	-	-	-	-	-	-	-	-	-	-	-	-	-	-	-	-	-	-	-	-	-	-	-	-	-	-	-	-	-	93	-	-	-	-	77	-	-	-	34	-	-	36	65	-	-	-
Temperature	TEMPC	-	-	-	-	-	-	100	-	-	-36	-35	-	-	-	-	-	-	-	-	-	-	-	-	-	-	-	-	-	-	-	-	-	-	32	-	-	-	-	-	-	-	-	-	-	-	-	-	-	-	-	-	-	-	-	-	-	-	-	-	-	-
Dissolved oxygen	DOX	-	-75	-	-	-	-	-	100	35	-	-	-	-	-48	-43	-	-	-	-	-	-	-	-	-38	-38	-	-	-36	-	-	-59	55	-	-	-	-31	-	-	-	55	-42	-	-34	56	-	-	-40	-	-	-	-45	36	-	-	-	-	-	-	-59	-46	-
Redox potential	EHmv	-	-64	-	-	-	-	-	35	100	-	-	-	-	-	-	-	-	-	-	-	-	-	-	-	-	-	-	-	-	-	-51	42	-	-	-	-	-	-	-	40	-65	-	-	-57	-	-	-	-	-	-	-	-	-	-	-	-	32	39	-52	-36	-
Land-surface altitude	LSALT	-	-	-	-	-	-	-36	-	-	100	92	30	-	-	-	-	-	-	-	-	-	-	-	-	-	-	-	-	-	-	-	-	-	-	-	-36	-	-	-	-	-	-	-	-	-	-	-	-	-	-	-	-	-	-	-	-	-	-	-	-	-
Well bottom altitude	WELLZALT	-	-	-	-	-	-	-35	-	-	92	100	-	-	-	-	-	-	-	-	-	-	-	-	-	-	-	-	-	-	-	-	-	-	-	-	-34	-	-	-	-	-	-	-	-	-	-	-	-	-	-	-	-	-	-	-	-	-	-	-	-	-
Well depth	WELLZ	31	-32	-	-	-	-	-	-	-	30	-	100	63	-	-	-	-	-	-	-	-	-	-	-	-	-	-	-	-	-	-	-	-	-	-	-	-	-	-	-	-	-	-	-38	-	-	-	-	-	-	-	-	-	-	-	-	-	-	-	-	-
Depth to water	WATRZ	-	-	-	-	-	-	-	-	-	-	-	63	100	-	-	-	-	-	-	-	-	-	-	-	-	-	-	-	-	-	-	-	-	-	-	-58	-	-	-	-	-51	-	-	-	-	-	-	-	-	-	-	-	-	-	-	-	-	-	-	-	-
pH, Field	pH	69	30	-33	-	-	-	-	-48	-	-	-	-	-	100	88	-	-	-	-	-37	-41	-	-	54	45	-	-	64	-	-	50	-50	-	-	-	32	-	-	-	-57	31	-35	66	-	-	-39	36	-	-	-	66	-	-	-	-	-	-54	-34	44	43	-
pH, Lab	pHL	69	-	-33	-	-	-	-	-43	-	-	-	-	-	88	100	-	-	-	-	-41	-42	-	-	60	44	-	-	66	-	-	50	-48	-	-	33	38	-	-31	-33	-50	-	-37	66	-	-	-35	36	-32	-	-	71	-	-	-	-	-	-53	-37	47	44	-
Specific conductance, lab	SCL	55	-	46	51	-	-	-	-	-	-	-	-	-	-	-	100	98	96	100	31	-	31	32	70	83	58	50	41	-	37	56	-36	41	-	-	-	-	-	-	-43	40	-	48	44	-	-	-	-	-	-	54	-	-	-	-	-	-55	-37	42	-	-
Specific conductance, field	SCF	53	-	47	53	-	-	-	-	-	-	-	-	-	-	-	98	100	95	98	33	-	32	-	67	83	58	52	41	-	40	53	-32	40	-	-	-	-	-	-	-42	41	-	46	43	-	-	-	-	-	-	52	-	-	-	-	-	-53	-34	38	-	-
Residue on evaporation 180C	ROE180	58	-	43	48	-	-	-	-	-	-	-	-	-	-	-	96	95	100	97	-	-	-	35	72	83	56	46	46	-	37	53	-33	42	-	-	-	-	-	-	-43	38	-	49	43	-	-	-	-	-	-	56	-	-	-	-	-	-54	-37	43	-	-
Total dissolved solids	TDS	57	-	48	48	-	-	-	-	-	-	-	-	-	-	-	100	98	97	100	-	-	-	32	73	84	56	48	43	-	38	58	-37	42	-	-	-	-	-	-	-44	43	-	49	45	-	-	-	-	-	-	55	-	-	-	-	-	-58	-39	45	-	-
Hardness	HRDcalc	-	-	79	-	-	-	-	-	-	-	-	-	-	-37	-41	33	33	-	-	100	98	88	-	-	-	-	-	-33	41	53	-	-	-	-	-	-	-	-	-	-	-	-	-	-	-	46	-	-	-	-	-31	-	-	-	-	-	-	34	-	-	-
Calcium	Ca	-35	-	72	-	-	-	-	-	-	-	-	-	-	-43	-42	-	-	-	-	98	100	77	-	-	-	-	-	-36	38	51	-	-	-	-	-	-	-	-	-	-	-	-	-33	-	-	46	-	-	-	-	-35	-	-	-	-	-	-	39	-	-	-
Magnesium	Mg	-	-	80	-	-	-	-	-	-	-	-	-	-	-	-	33	32	-	-	88	77	100	31	-	33	-	-	-	44	49	-	-	-	-	-	-	-	-	30	-	-	-	-	37	-	36	-	-	-	-	-	-	-	-	-	-	-	-	-	-	-

Parameter name	Parameter abbreviation	PC1	PC2	PC3	PC4	PC5	PC6	TEMPC	DOX	EHnw	LSALT	WELLZALT	WELLZ	WATRZ	pH	pHL	SCL	SCF	ROE180	TDS	HRDcalc	Ca	Mg	K	Na	ALK	Br	Cl	F	SiO_2
Potassium	K	40	-	-	-	71	-	-	-	-	-	-	-	-	-	-	32	-	35	32	-	-	31	100	38	32	-	-	-	31
Sodium	Na	79	30	-	37	-	-	-	-38	-	-	-	-	-	54	60	70	67	72	71	-	-	-	38	100	69	60	36	56	-
Alkalinity	ALK	72	32	36	-	-	-	-	-38	-	-	-	-	-	45	44	83	81	81	84	-	-	31	32	69	100	41	-	49	-
Bromide	Br	32	-	-	83	-	-	-	-	-	-	-	-	-	-	-	58	58	56	56	-	-	-	-	60	41	100	70	-	-
Chloride	Cl	-	-	-	89	-	-	-	-	-	-	-	-	-	-	-	50	52	46	48	-	-	-	-	36	-	70	100	-	-
Fluoride	F	66	-	-	-	-	-	-	-36	-	-	-	-	-	64	66	41	41	46	43	-33	-36	-	-	56	49	-	-	100	-
Silica	SiO_2	-	-	45	-	59	-	-	-	-	-	-	-	-	-	-	-	-	-	-	41	38	44	31	-	-	-	-	-	100
Sulfate	SO_4	-	-	85	-	-	-	-	-	-	-	-	-	-	-	-	37	40	37	38	53	51	49	-	-	-	-	-	-	31
Ammonia	NH_4N	55	70	-	-	-	-	-	-59	-51	-	-	-	-	50	50	56	53	53	58	-	-	-	35	68	63	41	33	58	-
Nitrate	NO_3N	-48	-66	-	-	-	-	-	55	42	-	-	-	-	-50	-48	-36	-32	-33	-37	-	-	-	-	-51	-54	-	-	-38	-
Phosphate	PO_4P	38	-	-	-	-	-	-	-	-	-	-	-	-	-	-	41	40	42	42	-	-	-	-	56	44	32	-	42	-
Total coliform	Tcoli	-	-	-	-	-	-	32	-	-	-	-	-	-	-	-	-	-	-	-	-	-	-	-	-	-	-	-	-	-
Aluminum	Al	-	-	-	-	-	-	-	-	-	-	-	-	-	-	31	-	-	-	-	-	-	-	-	-	-	-	-	-	-
Barium	Ba	-	-	-44	31	58	-	-	-31	-	-36	-34	-	-58	32	38	-	-	-	-	-	-	-	31	-	-	-	38	-	-
Beryllium	Be	-	-	-	-	-	-	-	-	-	-	-	-	-	-	-	-	-	-	-	-	-	-	-	-	-	-	-	-	-
Cadmium	Cd	-32	-	-	-	-	-	-	-	-	-	-	-	-	-	-31	-	-	-	-	-	-	-	-	-	-31	-	-	-33	-
Cobalt	Co	-	-	-	-	-	-	-	-	-	-	-	-	-	-	-33	-	-	-	-	-	-	30	-	-	-	-	-	-	-
Copper	Cu	-44	-57	-	-	-	-	-	55	40	-	-	-	-	-57	-50	-43	-42	-41	-44	-	-	-	-	-47	-53	-36	-	-57	-
Iron	Fe	-	76	-	-	-	-	-	-42	-45	-	-	-	-51	31	-	40	41	38	43	-	-	-	-	46	40	32	-	36	-
Lead	Pb	-32	-31	-	-	-	-	-	-	-	-	-	-	-	-35	-37	-	-	-	-	-	-	-	-	-34	-40	-	-	-40	-
Lithium	Li	85	-	-	-	42	-	-	-34	-	-	-	-	-	66	66	48	46	49	49	-	-33	-	52	78	63	45	-	60	-
Manganese	Mn	-	87	-	-	-	-	-	-56	-57	-	-	-38	-	-	-	44	43	43	45	-	-	37	-	34	45	-	-	-	-
Molybdenum	Mo	-	-	-	-	-	93	-	-	-	-	-	-	-	-	-	-	-	-	-	-	-	-	-	-	-	-	-	-	-

Parameter name	Parameter abbreviation	SO_4	NH_4N	NO_3N	PO_4P	Tcoli	Al	Ba	Be	Cd	Co	Cu	Fe	Pb	Li	Mn	Mo	Ni	Sr	Zn	Sb	As	B	Se	AlphaTh30	AlphaTh72	BetaCs30	BetaCs72	Rn222	U	Methane	Ethane	Propane
Potassium	K	-	35	-	-	-	-	31	-	-	-	-	-	-	52	-	-	-	58	-	-	-	43	-	-	-	44	58	-	-	34	-	-
Sodium	Na	-	68	-51	56	-	-	-	-	-	-	-47	46	-34	78	34	-	-	41	-	-	-	84	-	-	-	-	32	-59	-49	65	46	-
Alkalinity	ALK	-	63	-54	44	-	-	-	-	-31	-	-53	40	-40	61	45	-	-	34	-	-	-	69	-	-	-	-	-	-55	-31	46	36	-
Bromide	Br	-	41	-	32	-	-	-	-	-	-	-36	32	-	45	-	-	-	-	-	-	-	43	-	-	-	-	-	-	-	46	32	-
Chloride	Cl	-	33	-	-	-	-	38	-	-	-	-	-	-	-	-	-	-	-	-	-	-	-	-	-	-	-	31	-	-	33	-	-
Fluoride	F	-	58	-38	42	-	-	-	-	-33	-	-57	36	-40	60	-	-	-	-	-	-	-	64	-	-	-	-	-	-63	-58	44	46	-
Silica	SiO_2	31	-	-	-	-	-	-	-	-	-	-	-	-	-	-	-	-	42	-	-	-	-	-	-	-	-	-	-	-	-	-	-
Sulfate	SO_4	100	-	-	-	-	-	-52	-	-	-	-	-	-	-	-	-	43	-	-	-	-	-	-	-32	-	-	-	-	-	-	-39	-
Ammonia	NH_4N	-	100	-72	38	-	-	31	-	-	-	-67	71	-43	61	62	-	-	56	-	-36	-	72	-	-	-	30	49	-66	-62	83	54	-
Nitrate	NO_3N	-	-72	100	-	-	-	-	-	-	-	64	-52	39	-48	-55	-	-	-42	-	38	-	-63	45	-	-	-	-	46	47	-62	-41	-
Phosphate	PO_4P	-	38	-	100	-	-	-	-	-	-	-	36	-	-	-	-	-	-	-	-	-	38	-	-	-	-	-	-33	-	34	43	-
Total coliform	Tcoli	-	-	-	-	100	-	-	-	-	-	-	-	-	-	-	-	-	-	-	-	-	-	-	-	-	-	-	-	-	-	-	-
Aluminum	Al	-	-	-	-	-	100	-	35	-	-	-	-	-	-	-	-	-	-	-	31	-	-	-	-	-	-	-	-	-	-	-	-
Barium	Ba	-52	31	-	-	-	-	100	-	-	-	-	-	-	36	-	-	-	53	-	-	-	-	-	40	-	45	46	-	-	40	48	-
Beryllium	Be	-	-	-	-	-	35	-	100	-	32	-	-	-	-	-	-	-	-	-	-	-	-	-	-	-	-	-	-	-	-	-	-
Cadmium	Cd	-	-	-	-	-	-	-	-	100	-	-	-	-	-32	-	-	-	-	33	-	-	-33	-	-	-	-	-	-	-	-	-	-
Cobalt	Co	-	-	-	-	-	-	-	32	-	100	-	-	-	-	-	-	50	-	-	-	-	-	-	-	-	-	-	-	-	-	-	-
Copper	Cu	-	-67	64	-	-	-	-	-	-	-	100	-53	54	-44	-55	-	-	-40	-	41	-	-55	39	-	-	-	-32	56	50	-58	-39	-
Iron	Fe	-	71	-52	36	-	-	-	-	-	-	-53	100	-40	32	70	-	-	36	-	-	-	41	-	-	-	-	43	-55	-46	62	39	-
Lead	Pb	-	-43	39	-	-	-	-	-	-	-	54	-40	100	-32	-38	-	-	-	-	46	-	-38	35	-	-	-	-31	36	34	-40	-34	-
Lithium	Li	-	61	-48	-	-	-	36	-	-32	-	-44	32	-32	100	-	-	-	56	-	-	-	90	-	-	-	-	35	-51	-47	55	46	-
Manganese	Mn	-	62	-55	-	-	-	-	-	-	-	-55	70	-38	-	100	-	-	34	-	-	-	33	-46	-	-	-	40	-37	-	63	30	-
Molybdenum	Mo	-	-	-	-	-	-	-	-	-	-	-	-	-	-	-	100	-	-	-	-	63	-	-	-	36	-	-	-	55	-	-	-

[r-values multiplied by 100 and rounded; only values significant at $p < 0.01$ shown; PC, principle component; -, not significant; abbreviations are defined in Table 3.1].

Appendix 4

Distributions of continuous variables were compared among different sample classifications using notched box-plots (Velleman and Hoaglin, 1981; Helsel and Hirsch, 2002).

Where the median for a group is greater than the common reporting limit, it is displayed as a horizontal line within the box that is defined by the 25th and 75th percentiles for that group; otherwise, the median is displayed at the reporting limit. If the notched intervals around the medians for sample subsets do not overlap, the medians are statistically different at the 95-percent confidence interval.

References

Helsel, D. R., and Hirsch, R. M., 2002, Statistical methods in water resources: *U.S. Geological Survey Techniques of Water-Resources Investigations*, book 4, chap. A3, 510 p., accessed February 2019 at https://doi.org/10.3133/twri04A3.

McMahon, P.B., and Chapelle, F.H., 2008, *Redox processes and water quality of selected principal aquifer systems: Groundwater*, v. 46, no. 2, p. 259–271, accessed October 16, 2017, at http://onlinelibrary. wiley.com/doi/10.1111/j.1745- 6584.2007. 00385.x/abstract.

Pennsylvania Bureau of Topographic and Geologic Survey, 2001, *Bedrock geology of Pennsylvania: Pennsylvania Bureau of Topographic and Geologic Survey, Department of Conservation and Natural Resources*, digital data, accessed October 2017 at http://www.dcnr.state.pa.us/ topogeo/map1/ bedmap.aspx.

Velleman, P. F., and Hoaglin, D., 1981, *Applications, basics, and computing of exploratory data analysis:* Boston, Mass., Duxbury Press, 354 p.

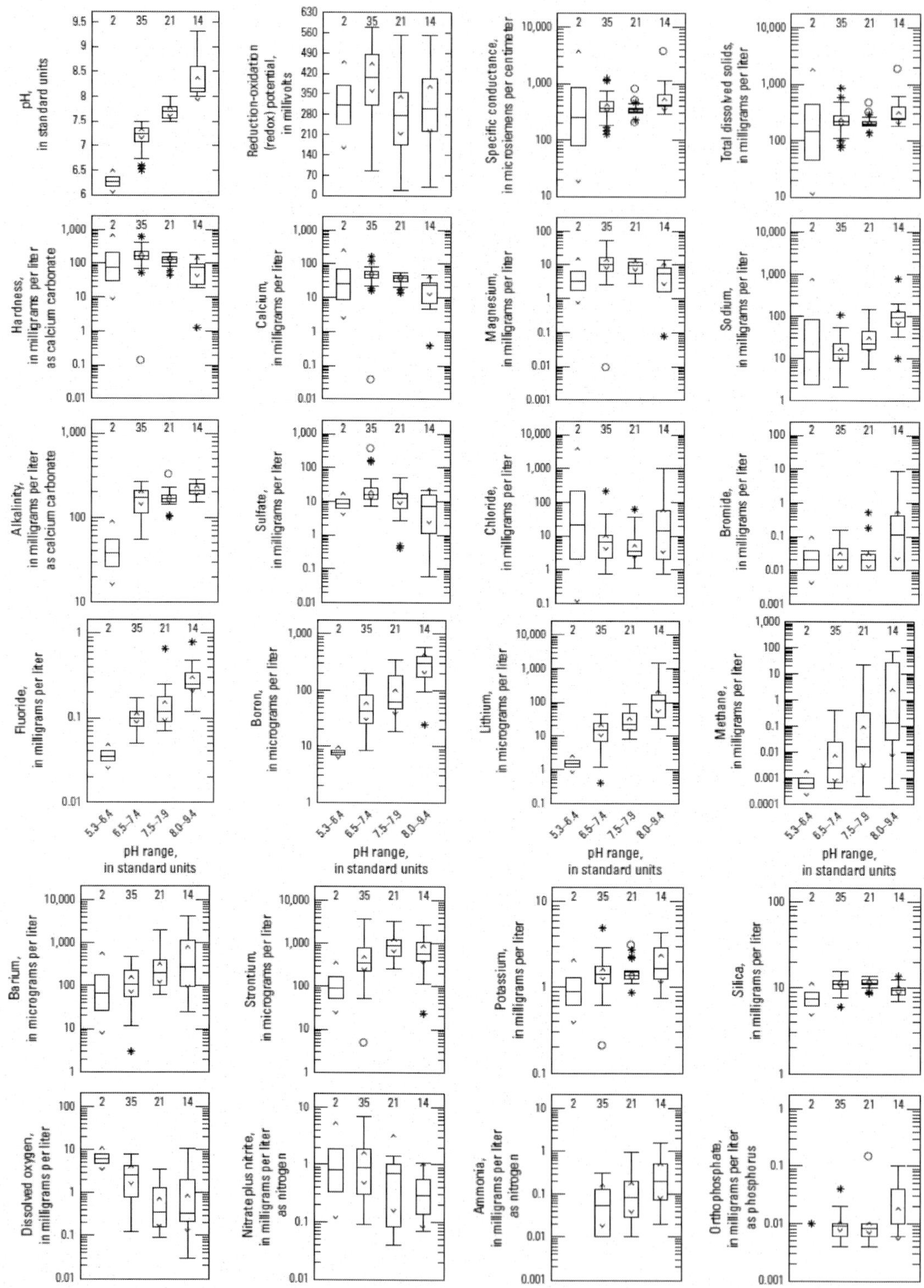

Figure 4.1. (Continued).

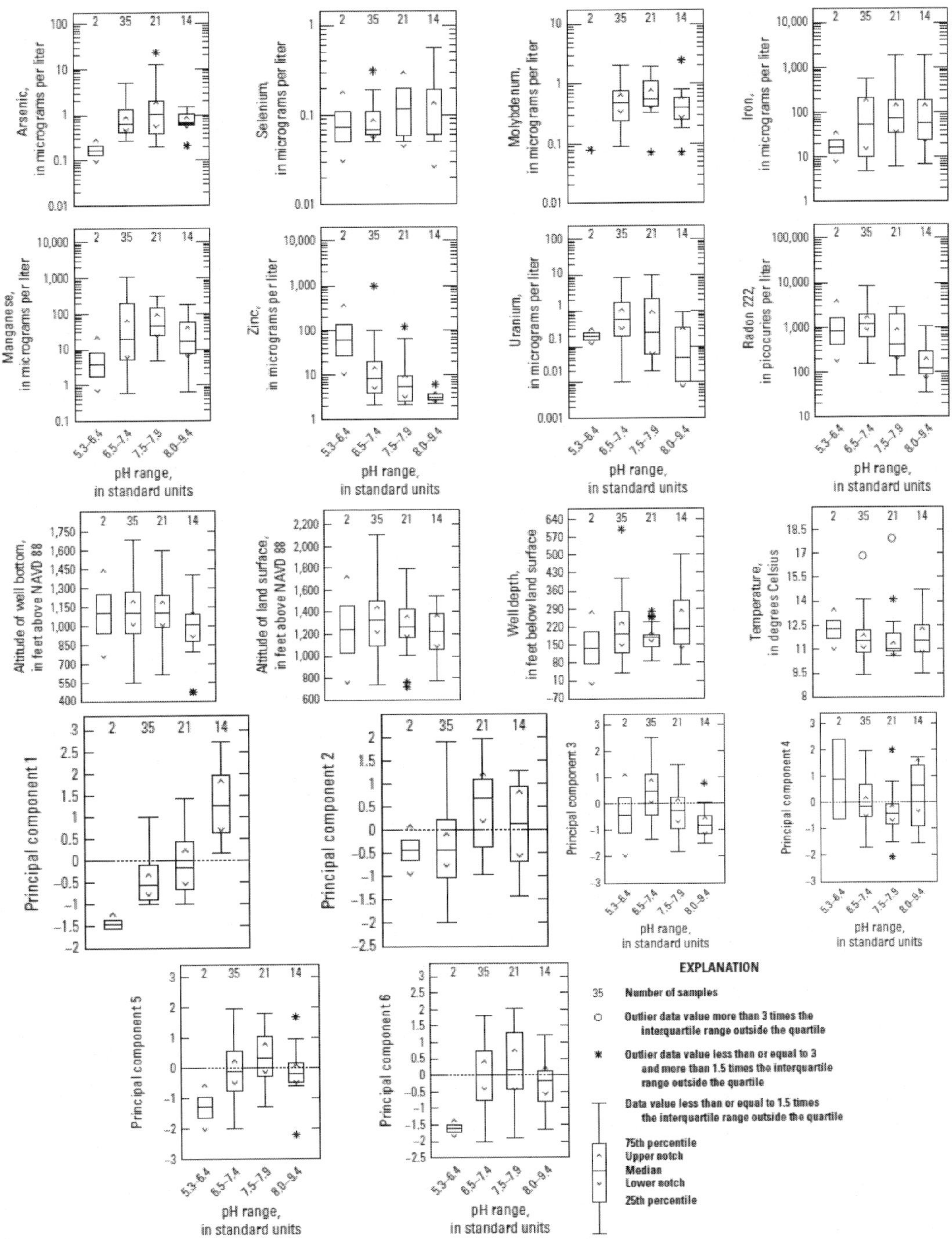

Figure 4.1. Composition of 72 groundwater samples from Bradford County, Pennsylvania, 2016, classified by pH class interval. The pH range is indicated on the x-axis (5.3 < pH < 6.4, n = 2; 6.5 < pH < 7.4, n = 35; 7.5 < pH < 7.9, n = 21; 8.0 < pH < 9.4, n = 14). (<, less than; n, number of samples; see table 3.1 for descriptions of principal components).

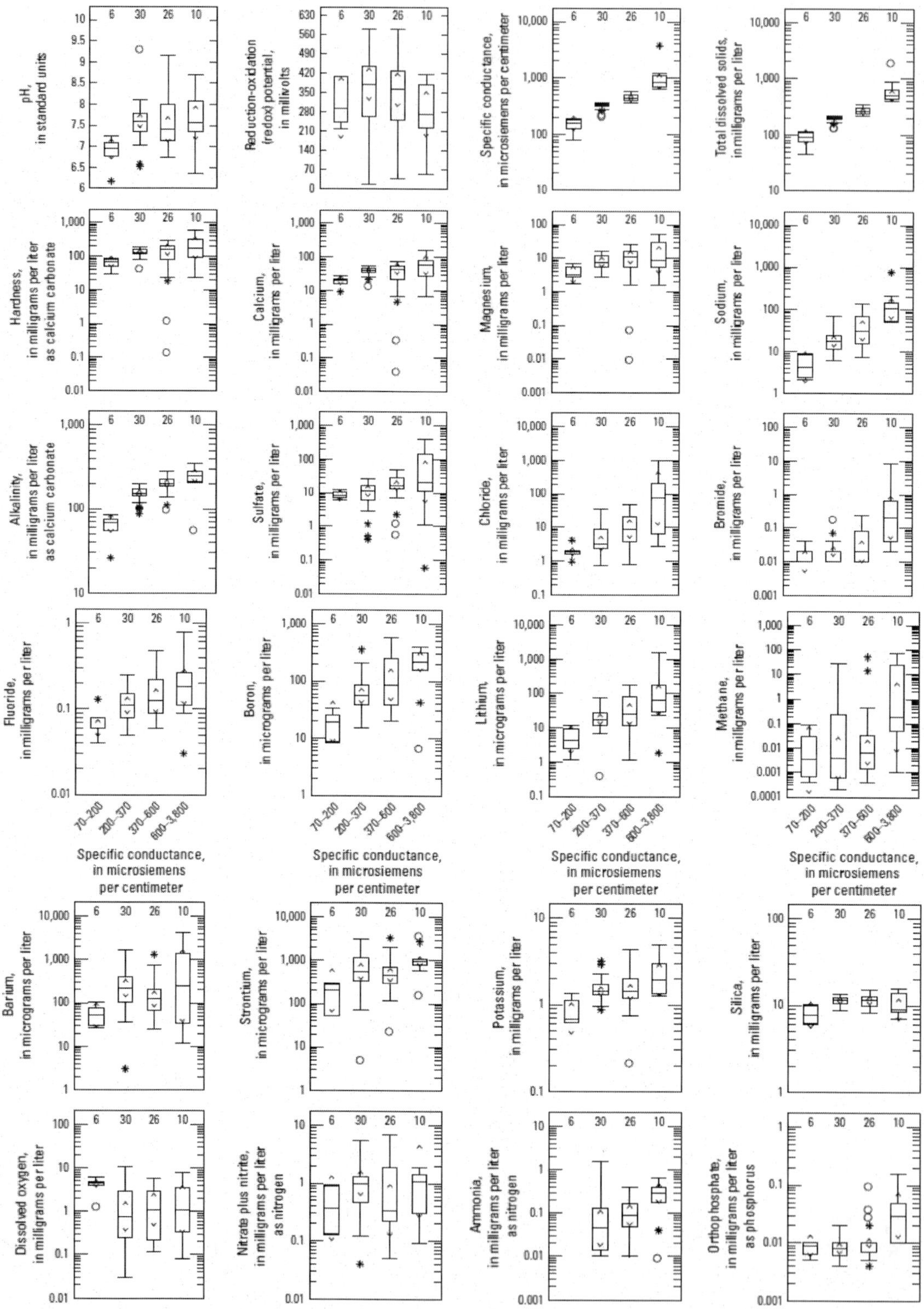

Figure 4.2. (Continued).

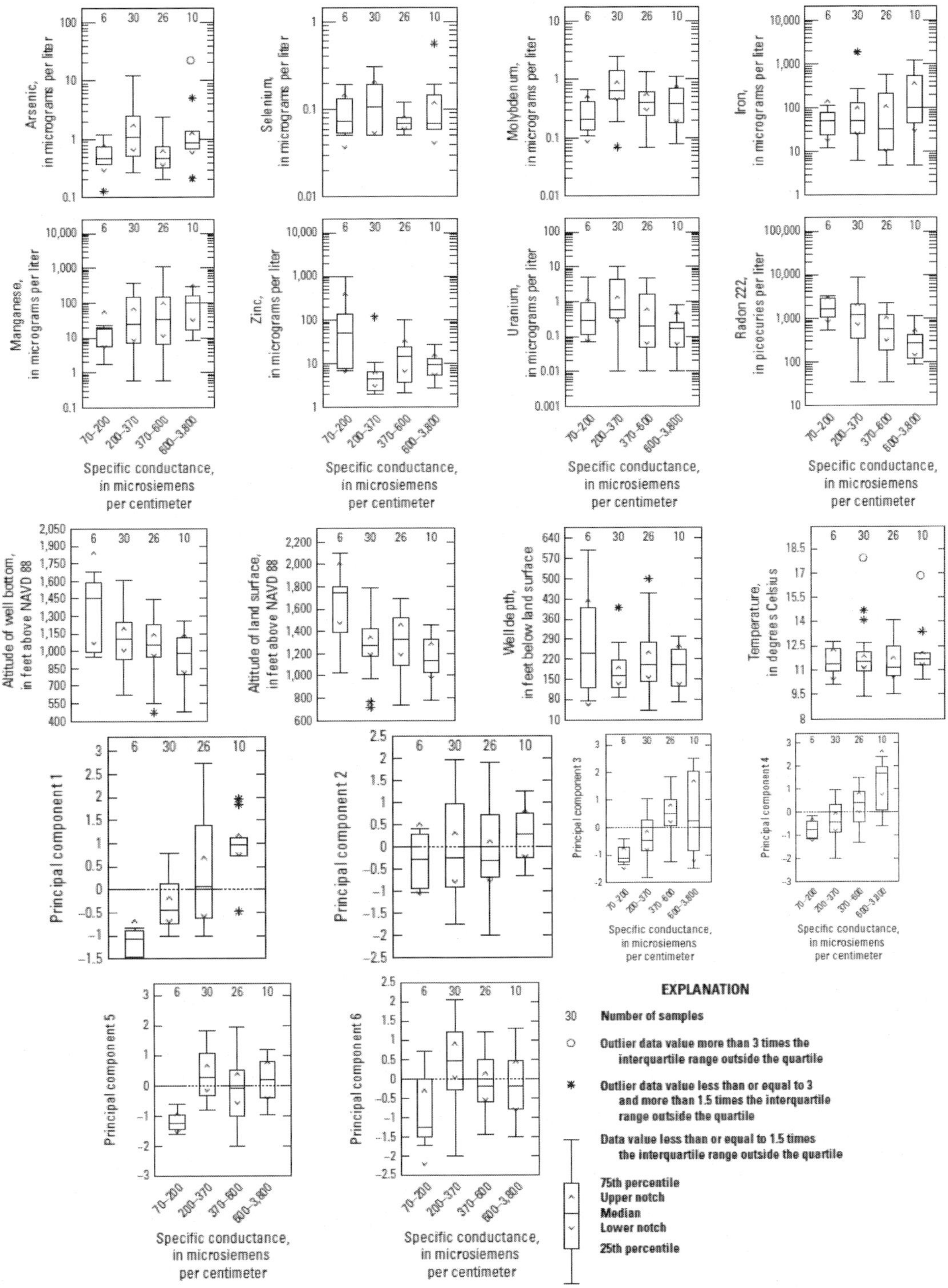

Figure 4.2. Composition of 72 groundwater samples from Bradford County, Pennsylvania, 2016, classified by specific conductance (SC) class interval. The SC range is indicated on the x-axis (70 < SC < 200 μS/cm, n = 6; 200 < SC < 370 μS/cm, n = 30; 370 < SC < 600 μS/cm, n = 26; 600 < SC < 3,800 μS/cm, n = 10). (<, less than; μS/cm, microsiemens per centimeter; n, number of samples; see table 3.1 for descriptions of principal components).

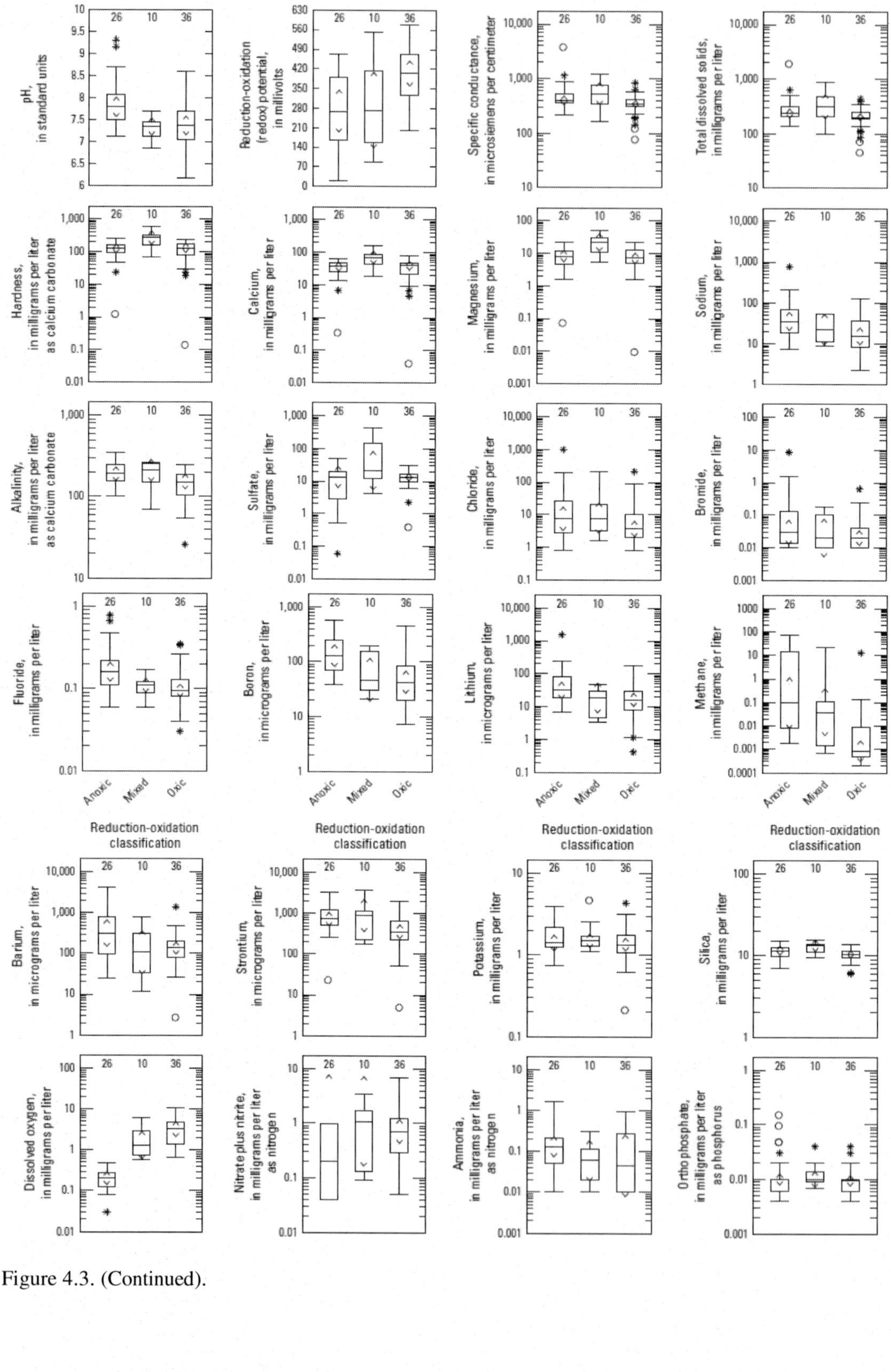

Figure 4.3. (Continued).

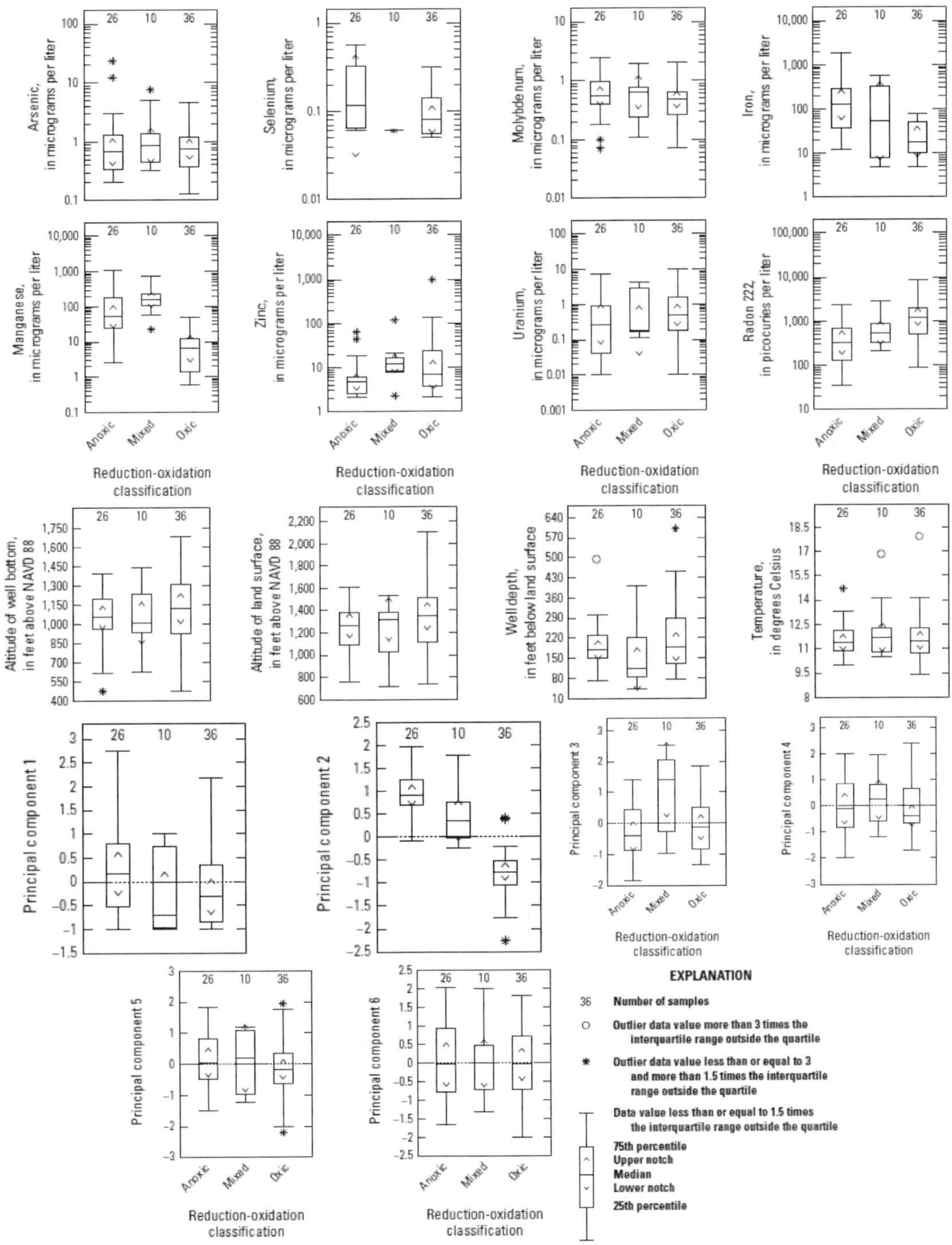

Figure 4.3. Composition of 72 groundwater samples from Bradford County, Pennsylvania, 2016, classified as anoxic (n = 26), mixed (n = 10), and oxic (n = 36) on the basis of dissolved oxygen concentration and other water-quality criteria of McMahon and Chapelle (2008). (n, number of samples; see Table 3.1 for descriptions of principal components).

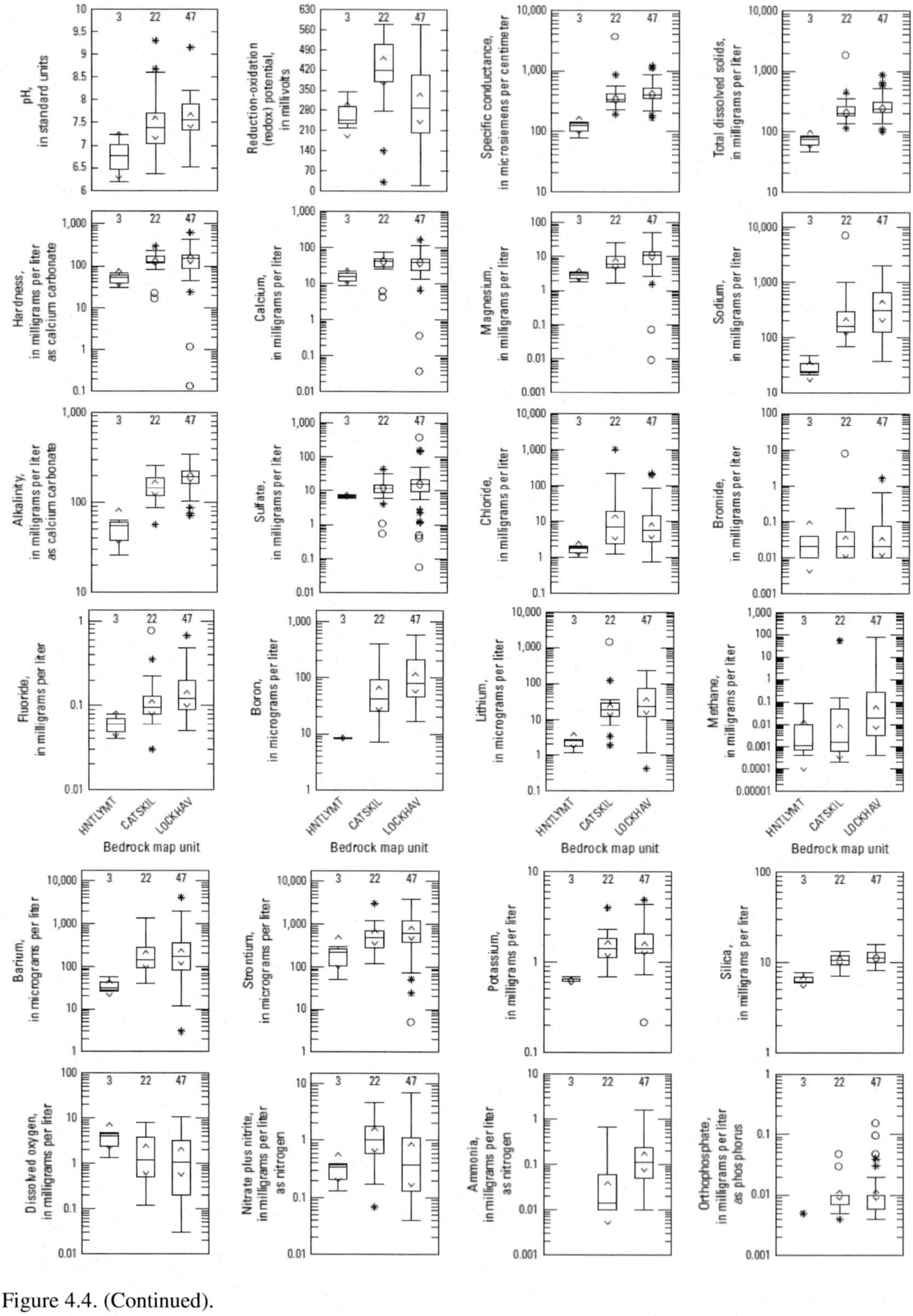

Figure 4.4. (Continued).

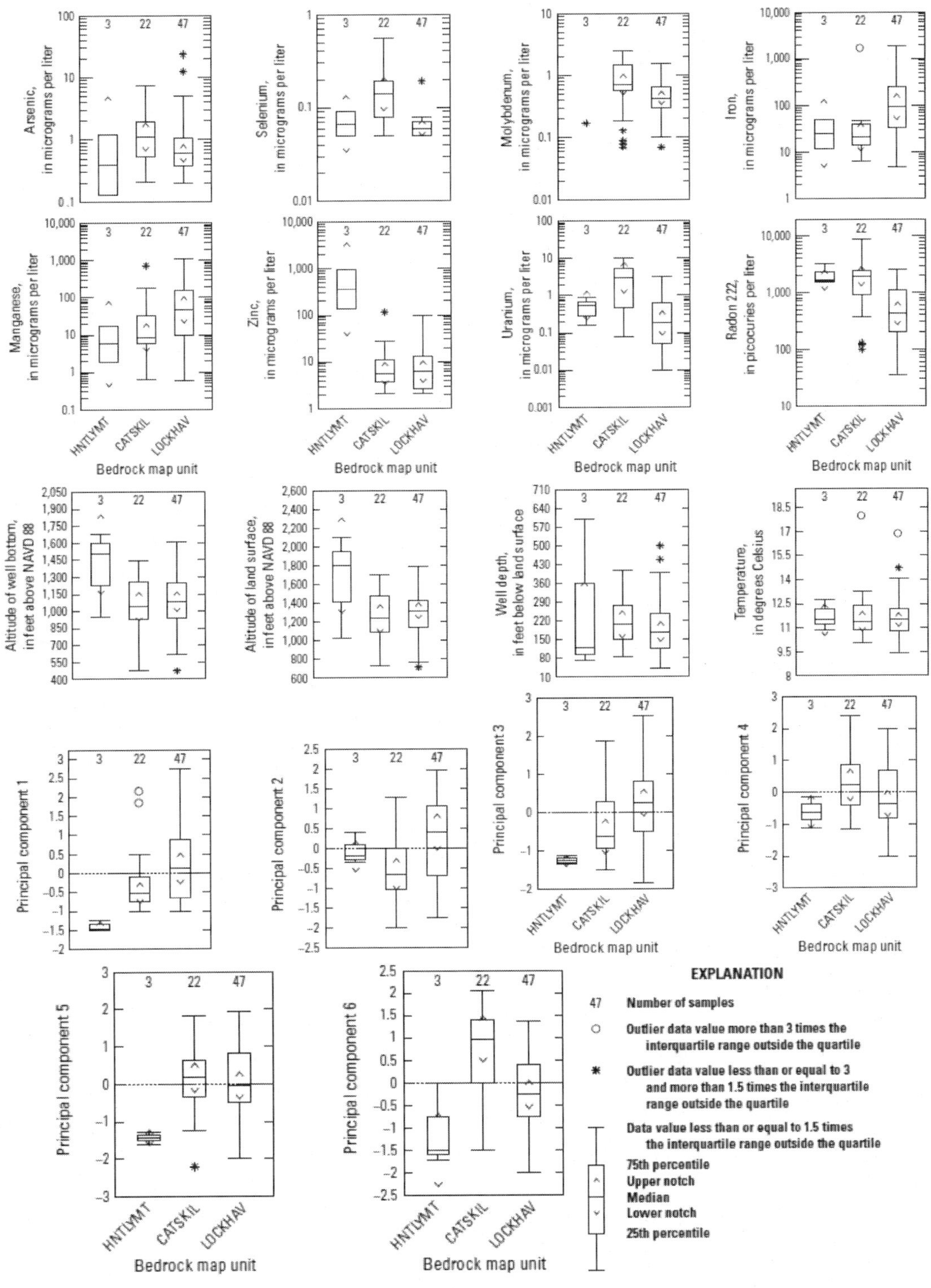

Figure 4.4. Composition of 72 groundwater samples from Bradford County, Pennsylvania, 2016, classified on the basis of bedrock map unit identified by Pennsylvania Bureau of Topographic and Geologic Survey (2001) as Huntley Mountain Formation (HNTLYMT; n = 3), Catskill Formation (CATSKIL; n = 22), and Lock Haven Formation (LOCKHAV; n = 47). (n, number of samples; see Table 3.1 for descriptions of principal components).

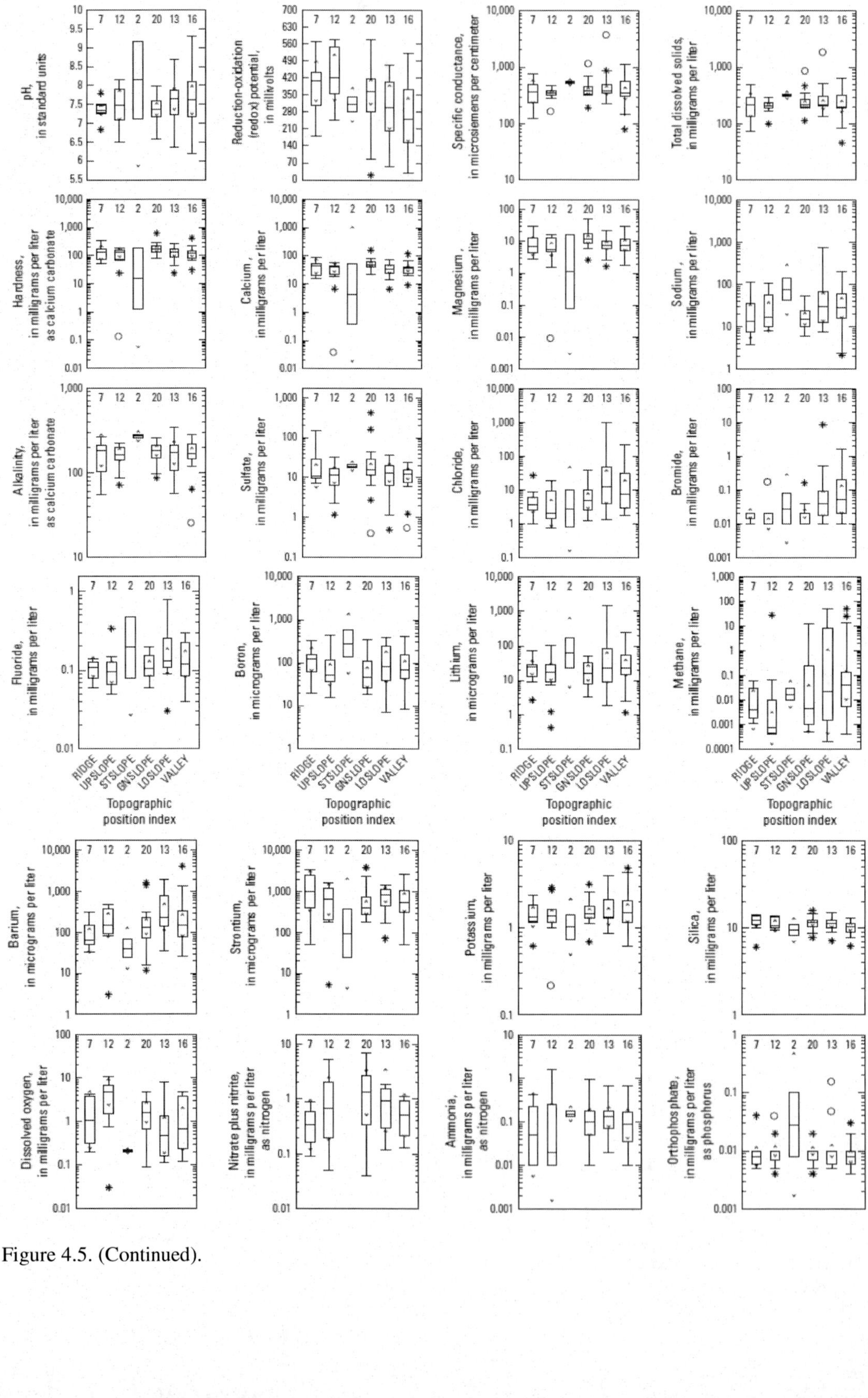

Figure 4.5. (Continued).

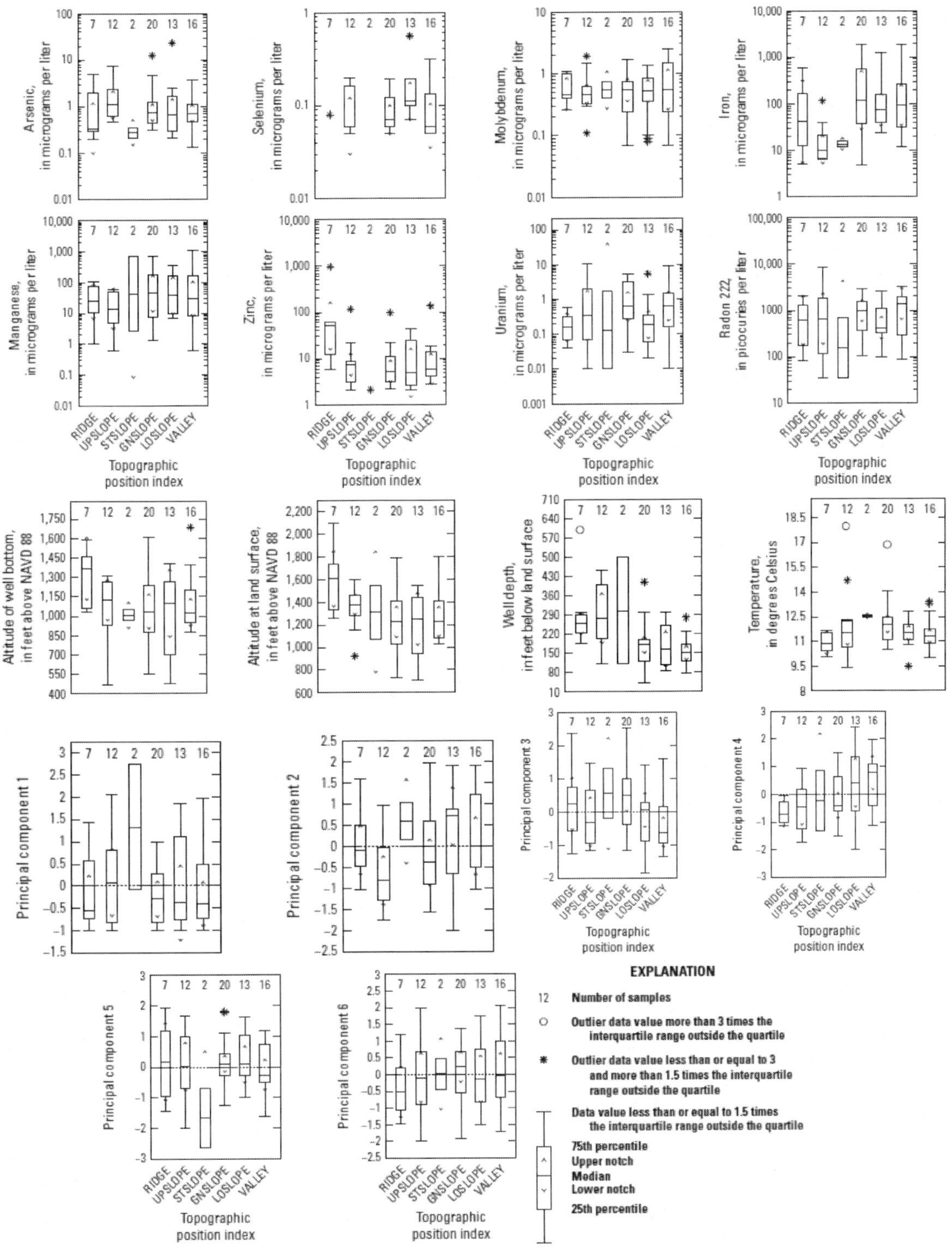

Figure 4.5. Composition of 72 groundwater samples from Bradford County, Pennsylvania, 2016, classified on the basis of topographic position index (TPI). The TPI classes are ridge (RIDGE; n = 7), upper slope (UPSLOPE; n = 12), steep slope (STSLOPE; n = 2), gentle slope (GNSLOPE; n = 20), lower slope (LOSLOPE; n = 13), and valley (VALLEY; n = 16). (n, number of samples; see Table 3.1 for descriptions of principal components).

In: Drinking Water
Editor: Erwin Jon Orosz

ISBN: 978-1-53617-284-3

Chapter 3

ARSENIC AND URANIUM IN PRIVATE WELLS IN CONNECTICUT, 2013–15*

Sarah M. Flanagan and Craig J. Brown

MAJOR FINDINGS

- Nearly 1 out of 15 (7 percent) water samples from 674 private wells tested in Connecticut contained either arsenic or uranium at concentrations that exceed the U.S. Environmental Protection Agency's (EPA) maximum contaminant levels (MCLs) enforceable for drinking-water supplies of 10 micrograms per liter (μg/L) for arsenic or 30 μg/L for uranium.
- Of the 81 geologic units studied, 19 had at least one sample with arsenic or uranium concentrations that exceeded the MCL.

INTRODUCTION

The occurrence of arsenic and uranium in groundwater at concentrations that exceed drinking-water standards is a concern because of the potential adverse effects on human health. Some early studies of arsenic occurrence in groundwater considered anthropogenic causes, but more recent studies have focused on sources of naturally occurring arsenic to groundwater, such as minerals within aquifer materials that are in contact with groundwater. Arsenic and uranium in groundwater in New England have

* This is an edited, reformatted and augmented version of Open-File Report of U.S. Department of the Interior, U.S. Geological Survey, prepared in cooperation with the Connecticut Department of Public Health No. 2017–1046, dated May 2017.

been shown to have a strong association to the geologic setting (Ayotte and others, 2003, 2006) and nearby streambed sediment concentrations (Robinson and Ayotte, 2006). In New Hampshire and Massachusetts, arsenic and uranium concentrations greater than human-health benchmarks have shown distinct spatial patterns when related to the bedrock units mapped at the local scale (Montgomery and others, 2003; Colman, 2011; Flanagan and others, 2014).

The Connecticut Department of Public Health (DPH; 2016) reported that there are about 322,600 private wells in Connecticut serving approximately 823,000 people, or 23 percent of the State's population. The State does not require that existing private wells be routinely tested for arsenic, uranium, or other contaminants; consequently, private wells are only sampled at the well owner's discretion or when they are newly constructed. The U.S. Geological Survey (USGS), in cooperation with the DPH, completed an assessment in 2016 on the distribution of concentrations of arsenic and uranium in groundwater from bedrock in Connecticut (Figure 1). This report presents the major findings for arsenic and uranium concentrations from water samples collected from 2013 to 2015 from private wells.

Sources of Data on Arsenic and Uranium Concentrations

The main objective of this study was to compile and analyze arsenic and uranium concentrations from private wells throughout Connecticut (Figure 1). In cooperation with the DPH, local health departments and districts used volunteers to visit randomly selected houses in their districts and collect an unfiltered water sample from an untreated source in the home. Staff from DPH also distributed bottles to homeowners at four agricultural fairs in Bethlehem, Durham, Goshen, and Woodstock; at water fair events in East Hampton and Colchester; and one home show in Hartford. These homeowners were instructed to collect an unfiltered water sample prior to any existing treatment system. The water samples, collected from 2013 to 2015, were submitted to the DPH Laboratory in Hartford for trace-element analysis according to EPA method 200.8 (U.S. Environmental Protection Agency, 1994).

The towns of Stamford, Weston, and Wilton have substantially more water samples and wells than the other studied towns. On average, 7 wells were sampled in most towns, whereas, 81 wells were sampled in Wilton; 110 wells, in Weston; and 732 wells, in Stamford, resulting in an unbalanced distribution of water samples in the State. To balance the distribution for the purposes of this study, 7 to 14 wells from Stamford, Weston, and Wilton were randomly selected for inclusion in the statewide dataset. This subset of samples represented less than 5 percent of the final statewide dataset, which consisted of 660 arsenic samples and 589 uranium samples collected from 674 wells. Data for these 674 wells are available in Flanagan and Brown (2017). All available data from these three towns are discussed in the "Comparison of Arsenic and Uranium Exceedance Rates in Three Towns" section.

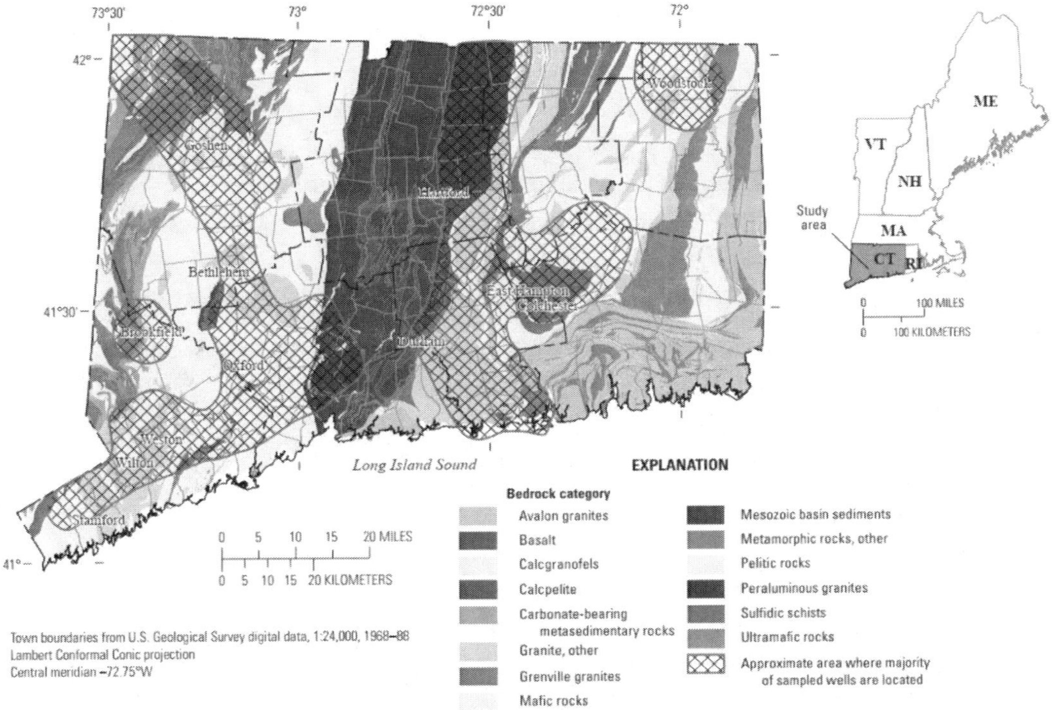

Figure 1. Locations of major categories of bedrock in Connecticut and areas where the majority of sampled wells are located. Modified bedrock categories are described in Robinson and Kapo (2003).

Table 1. Inventory of standard reference samples for arsenic and uranium concentrations, in 2015

SRS lot#	Submission Date	Arsenic concentration, in µg/L			Uranium concentration, in µg/L		
		MPV	DPH	RPD[1]	MPV	DPH	RPD[1]
T–219	6/30/2015	3.51	<3.00	na	1.58	1.50	5.2
T–201	7/28/2015	24.4	23.0	5.9	9.22	9.20	0.2
T–201b	8/18/2015	24.4	23.0	5.9	9.22	9.10	1.3
T–217A	8/31/2015	5.99	5.80	3.2	1.78	1.70	4.6
T–217B	9/29/2015	5.99	5.20	14.1	1.78	1.70	4.6
T–201c	9/29/2015	24.4	22.0	10.3	9.22	9.30	0.8

[SRS, standard reference sample; #, number; µg/L, microgram per liter; MPV, most probable value; DPH, Connecticut Department of Public Health Laboratory in Hartford, Conn.; <, less than; na, not applicable].

[1]Relative percent difference = [|(Sample 1 – Sample 2)| / Average (Sample 1 + Sample 2)] × 100, where sample 1 is the MPV value and sample 2 is the DPH value.

As part of the quality assurance plan for this study, the USGS submitted six standard reference samples to the DPH Laboratory with concentrations of arsenic and uranium spanning the range of expected environmental values. These samples were supplied by the USGS Branch of Quality Systems (U.S. Geological Survey, 2016) in Denver, Colo., and provided an independent analysis of measurable bias. Results from the analysis of the standard reference samples (Table 1) indicated no measurable bias.

ARSENIC AND URANIUM CONCENTRATIONS IN THE STATE

Arsenic concentrations ranged mostly (95th percentile) from less than (<) 3 to 7.1 micrograms per liter (µg/L). Uranium concentrations ranged mostly (95th percentile) from <1 to 23 µg/L (Table 2). Arsenic at concentrations at or above the minimum reporting level (MRL) of 3 µg/L was measured in 9.1 percent of samples. Uranium at concentrations at or above the MRL of 1 µg/L was measured in 42.1 percent of the samples. Statewide, 3.6 percent of samples had arsenic concentrations that exceeded the MCL of 10 µg/L, and 3.9 percent had uranium concentrations that exceeded the MCL of 30 µg/L (Table 2). Overall, about 7 percent of the samples had concentrations of arsenic or uranium that exceeded an MCL.

Table 2. Arsenic and uranium concentrations in water samples from 674 private wells in Connecticut, 2013–15

Trace element	No. of samples	MRL, in µg/L	Percentage of samples with concentrations equal to or exceeding MRL	Concentration, in µg/L						MCL, in µg/L	Percentage of samples with concentrations exceeding MCL
				Min.	Percentile				Max.		
					50	75	90	95			
Arsenic	660	3	9.1	<3	<3	<3	<3	7.1	470	10	3.6
Uranium	589	1	42.1	<1	<1	3.2	9.3	23	766	30	3.9

[No., number; MRL, minimum reporting level; µg/L, microgram per liter; MCL, U.S. Environmental Protection Agency maximum contaminant level for public water supplies; Min, minimum; Max, maximum; <, less than].

ARSENIC AND URANIUM OCCURRENCE IN RELATION TO BEDROCK GEOLOGY

There are distinct areas in Connecticut where one or more groundwater samples have arsenic and uranium concentrations greater than their respective MCLs (Figure 2). Data were grouped in relation to mapped bedrock units (referred to as geologic units in this report) identified on the bedrock geological map of Connecticut (Rodgers, 1985). It was assumed that each well was drilled and completed in the geologic unit represented at thewell's location on the geologic map. Geologic units are rocks of a specific geologic age that have unique mineral and physical characteristics, varying degrees of resistance to weathering, and similar processes of rock formation. Arsenic and uranium samples were grouped according to the geologic units in which the sampled wells are located (Table 3).

Of the 156 geologic units in the State, 81 units (covering 82.6 percent of the land area), were represented by at least one water sample analyzed for arsenic and (or) uranium (Table 3). Twenty-one geologic units had only 1 sample, 43 geologic units had 2 to 10 samples, and 17 geologic units had more than 10 samples. The 81 geologic units were organized under 10 different bedrock categories and are listed in Table 3. These bedrock categories (Figure 1) are based on groups of individual geologic units with similar geochemical and lithological properties (Robinson and Kapo, 2003). The percentage of samples in each geologic unit with arsenic and uranium concentrations that exceeded MCLs was computed and then geologic units were grouped and colored based on percentage ranges (Figure 2). The exceedance percentages for geologic units computed for this study may not represent the actual hazard for existing and future wells in these units. Nonetheless, this report provides new information on arsenic and uranium contamination at the State scale.

Results of this analysis indicate that the geologic units were markedly different in terms of arsenic and uranium concentrations that exceeded MCLs (Table 3). Nine of 81 geologic units had at least one sample with arsenic concentrations that exceeded the MCL of 10 µg/L. Fourteen geologic units had at least one sample with uranium concentrations that exceeded the MCL of 30 µg/L. None of the geologic units in the carbonate-bearing metasedimentary rocks, Grenville granites, mafic rocks, or metamorphic rocks, other bedrock categories had arsenic concentrations that exceeded the MCL (Table 3). None of the geologic units in the Avalon granites, carbonate-bearing metasedimentary rocks, Grenville granites, Mesozoic basin sediments, or sulfidic schists bedrock categories had uranium concentrations that exceeded the MCL. The pelitic rocks bedrock category had three geologic units with at least one arsenic concentration that exceeded the MCL, and the granite, other bedrock category had seven geologic units with at least one uranium concentration that exceeded the MCL—the most among the 10 bedrock categories.

In the greater than (>) 20 to 30 percent group (Table 3), the Taine Mountain and Collinsville Formation undivided (4 samples) and the Glastonbury Gneiss (20 samples) were the only geologic units with arsenic and (or) uranium concentrations that exceeded MCLs. In the >30 percent group, the Carringtons Pond Member of the Trap Falls Formation (3 samples) was the only geologic unit with concentrations that exceeded the MCL for arsenic, and the Dalton Formation and the Harrison Gneiss (3 samples each) were the only geologic units with concentrations that exceeded the MCL for uranium (Table 3).

The well with the highest arsenic concentration (470 µg/L), from the town of Oxford, is completed in The Straits Schist geologic unit of the Pelitic rocks bedrock category. Although 8.3 percent of the 12 samples in this geologic unit also have high arsenic (>10 µg/L), six other geologic units (Table 3) in the State have higher percentages of samples with high arsenic. This finding shows that wells with elevated arsenic are not always in the most high-risk areas.

The well with the highest uranium concentration (766 μg/L), from the town of Brookfield, is completed in the Dalton Formation geologic unit of the Metamorphic rock, other bedrock category. In a neurotoxicity case study in Brookfield, it was determined that a family was unknowingly exposed to well water containing 866 to 1,166 μg/L of uranium (Magdo and others, 2007). The private well in the case study was located in the Brookfield Gneiss geologic unit, a mafic rock common throughout the Appalachian ridges of western Connecticut. Magdo and others (2007) sampled 10 other wells in close proximity to the case-study well and discovered widely variable uranium concentrations ranging from 0.21 to 521 μg/L. Altogether, 4 of the 11 wells in Magdo and others (2007) exceeded the MCL for uranium. However, none of the 12 water samples from the Brookfield Gneiss geologic unit of the Mafic rocks bedrock category in this study exceeded the MCL for uranium (Table 3). This finding shows that uranium concentrations can be highly variable, even in wells in close proximity to each other or in similar geologic settings. These two examples for wells with elevated arsenic and uranium concentrations highlight the importance of individual well testing for naturally occurring contaminants.

COMPARISON OF ARSENIC AND URANIUM EXCEEDANCE RATES IN THREE TOWNS

A select group of geologic units have concentrations of arsenic and (or) uranium that exceed MCLs (Figure 2; Table 3). Therefore, a town's vulnerability to arsenic and uranium contamination may depend greatly on the extent to which these geologic units are within its borders. This is demonstrated by examining all available water-quality data for three towns with the largest number of sampled wells (Table 4). In Stamford, which is in southwestern Connecticut (Figure 2), 0.1 percent of 732 samples had high arsenic concentrations (>10 μg/L) and 11.2 percent of 731 samples had high uranium concentrations (>30 μg/L; Table 4). This difference in occurrence rates between the two elements is likely explained by the geologic setting of the town. Much of Stamford is underlain by the Harrison Gneiss geologic unit of the Mafic rocks bedrock category and the Trap Falls Formation and Ordovician granitic gneiss undivided geologic unit of the Granite, other bedrock category. For these two units, none of the samples in the statewide dataset had high arsenic concentrations, yet 33.3 and 11.1 percent, respectively, of samples had high uranium concentrations (Table 3).

The towns of Weston and Wilton are adjacent to each other in southwestern Connecticut (Figures 1 and 2). However, more than 47 percent of samples in Weston had high arsenic concentrations, but only 2.5 percent of samples in Wilton had high arsenic concentrations (Table 4). In Weston, the dominant geologic unit is the Ordovician granitic gneiss of the Granite, other bedrock category, and 14.3 percent of 56 samples from this unit in the statewide dataset had high arsenic concentrations (Table 3). In Wilton, the dominant geologic units are the Harrison Gneiss and the Pumpkin Ground Member of Harrison Gneiss (Mafic rocks bedrock category), and the Trap Falls Formation and Ordovician granitic gneiss undivided of the Granite, other bedrock category; none of the wells from these units in the statewide dataset had high arsenic concentrations (Table 3).

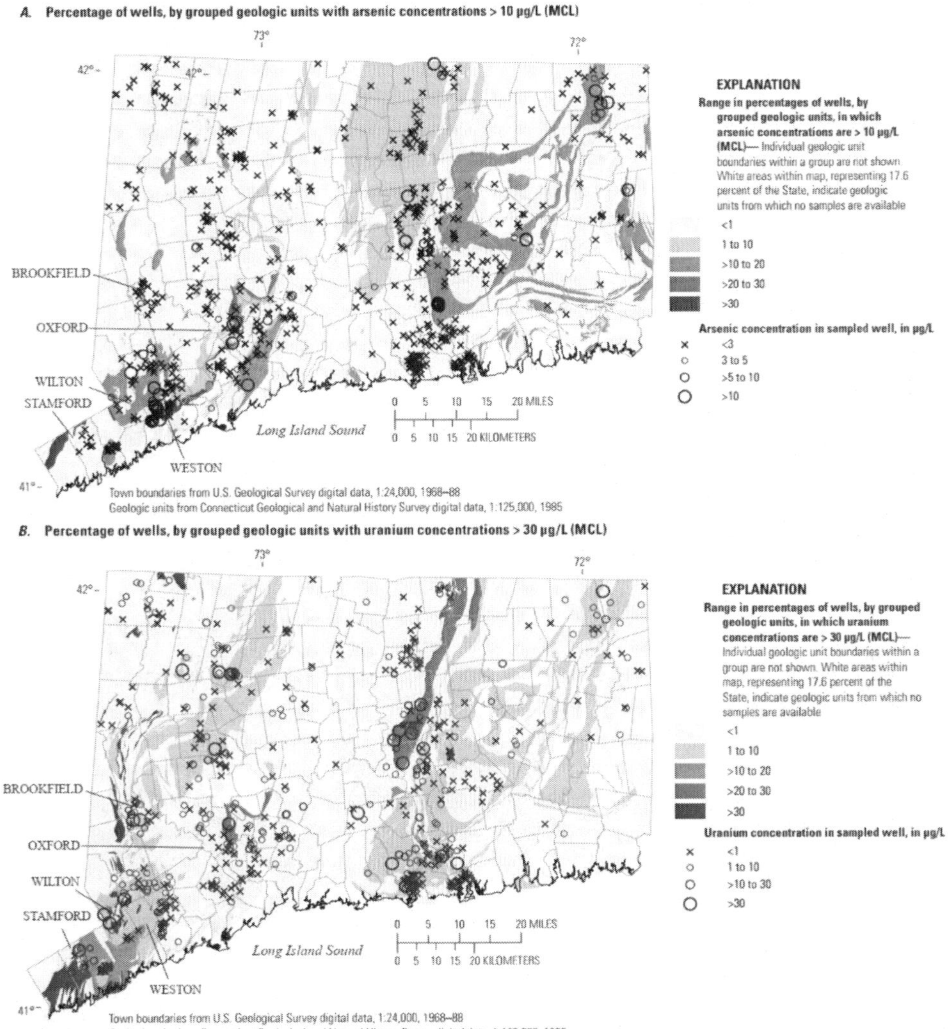

Figure 2. Range of percentages of private wells in Connecticut, by grouped geologic units, in which *A*, arsenic concentrations exceeded the maximum concentration level (MCL) of 10 micrograms per liter (µg/L) and *B*, uranium concentrations exceeded the MCL of 30 µg/L. Concentrations of arsenic and uranium in water samples collected from 674 individual wells and the spatial distribution of the wells are also shown. Well locations have been offset by one-fourth mile to maintain the confidentiality of the well owner's identity. Geologic units are listed in table 3. See Rodgers (1985) for the location and description of individual geologic units. <, less than; > greater than.

Table 3. Arsenic and uranium concentrations that exceed maximum contaminant levels from 674 private wells in Connecticut, by geologic unit and major bedrock category, 2013–15

Geologic unit name	Geologic unit code	Number of samples		MCL, percentage1 of water samples with concentrations, in micrograms per liter		Percentage of study area underlain by geologic unit2
		Arsenic	Uranium	Arsenic >10 µg/L	Uranium >30 µg/L	
Avalon granite						
"Scituate" Granite Gneiss	Zss	1	1	0	0	0.7
Hope Valley Alaskite Gneiss	Zsh	5	3	20.0	0	2.1
Plainfeld Formation	Zp	3	2	0	0	1.4
porphyritic phase of Potter Hill Granite Gneiss	Zspp	1	1	0	0	<0.2
Potter Hill Granite Gneiss	Zsph	2	2	0	0	1.3
Potter Hill Granite Gneiss and Narragansett Pier Granite undi vided	Zsph + Pn	1	1	0	0	<0.2
Rope Ferry Gneiss of the "Waterford Group"	Zwr	5	5	0	0	1.1
Calcgranofels						
Fly Pond (calc-silicate) member of Tatnic Hill Formation	Otaf	1	—	0	—	0.4
Hebron Gneiss	SOh	45	24	17.8	4.2	4.7
lower member of Bigelow Brook Formation	SObl	2	2	0	0	0.6
Southbridge Formation	SOs	5	2	0	0	1.0
Carbonate-bearing metasedimentary rocks						
basal marble member of Walloomsac Schist	Owm	4	4	0	0	0.5
Stockbridge Marble	OCs	3	2	0	0	0.8
unit b of Stockbridge Marble	Csb	6	6	0	0	0.4
unit c of Stockbridge Marble	Csc	3	3	0	0	0.5
units e and d of Stockbridge Marble	Ose	1	1	0	0	<0.2
Granite, other						
"Eastford gneiss phase" of Canterbury Gneiss	Dce	2	1	0	0	0.5
Canterbury Gneiss	Dc	2	1	0	0	1.2
Glastonbury Gneiss	Ogl	22	20	4.5	25.0	1.7
lower member of Middletown Formation	Oml	3	3	0	0	<0.2
Middletown Formation	Om	10	11	0	0	1.0
Monson Gneiss	Omo	47	45	0	4.4	2.4
Nonewaug Granite	Dng	10	10	0	10.0	0.6
Ordovician granitic gneiss	Og	56	41	14.3	9.8	2.1

Geologic unit name	**Geologic unit code**	**Number of samples**		**MCL, percentage1 of water samples with concentrations, in micrograms per liter**		**Percentage of study area underlain by geologic unit2**
		Arsenic	Uranium	Arsenic >10 μg/L	Uranium >30 μg/L	
Trap Falls Formation and Ordovician granitic gneiss undivided	Otf + Og	3	9	0	11.1	1.0
upper member of Middletown Formation	Omu	9	9	0	11.1	0.2
Waterbury Gneiss	Cwb	1	—	0	—	0.8
Waterford Group	Zw	26	25	0	4.0	0.8
Grenville granite						
Gneiss of Highlands massifs	Yg	4	4	0	0	0.6
pink granitic gneiss	Ygr	2	2	0	0	1.1
Mafic rocks						
amphibolite-bearing unit of Manhattan Schist	Cma	3	3	0	0	0.4
Beardsley Member of Harrison Gneiss	Ohb	9	8	0	0	0.5
Brookfeld Gneiss	Ob	14	12	0	0	1.2
dioritic phase of Lebanon Gabbro	Dld	1	1	0	0	<0.2
gneiss (metavolcanic) member of Brimfeld Schist	Obrg	5	5	0	0	0.3
Harrison Gneiss	Oh	3	3	0	33.3	0.9
hornblende gneiss member of Collinsville Formation	Ocg	3	3	0	0	0.4
Lebanon Gabbro	Dl	2	2	0	0	0.3
massive mafc rock in Middletown Formation	Omm	1	1	0	0	<0.2
Pumpkin Ground Member of Harrison Gneiss	Ohp	4	4	0	0	0.4
Quinebaug Formation	Oq	11	5	0	0	1.8
Mesozoic basin sediments						
East Berlin Formation	Jeb	5	4	0	0	1.2
Hampden Basalt	Jha	1	1	0	0	0.3
New Haven Arkose	TRnh	20	20	0	0	5.7
Portland Arkose	Jp	43	41	4.7	0	7.3
Shuttle Meadow Formation	Jsm	1	1	0	0	0.4
Metamorphic rocks, other						
Bristol Gneiss	Obs	2	2	0	0	0.4
Clough Quartzite	Sbc	1	1	0	0	<0.2
Dalton Formation	Cd	3	3	0	33.3	0.7
Fitch Formation	Sbf	1	1	0	0	<0.2
hornblende gneiss and amphibolite	Ygh	1	1	0	0	0.7
Llayered gneiss	Ygn	3	3	0	0	1.7
quartzite unit in Plainfeld Formation	Zpq	1	1	0	0	0.6
Tatnic Hill Formation	Ota	22	20	0	5.0	2.8

Table 3. (Continued)

Geologic unit name	Geologic unit code	Number of samples		MCL, percentage1 of water samples with concentrations, in micrograms per liter		Percentage of study area underlain by geologic unit2
		Arsenic	Uranium	Arsenic >10 µg/L	Uranium >30 µg/L	
Pelitic rocks						
amphibolite unit in Ratlum Mountain Schist	Ora	1	1	0	0	<0.2
basal member of Taine Mountain Formation around Waterbury dome	Otb	2	2	0	0	0.4
Collins Hill Formation	Och	4	4	0	0	0.8
Collinsville Formation	Oc	20	20	0	0	1.0
Golden Hill Schist	Ogh	3	3	0	0	0.3
Hoosac Schist	Ch	10	10	0	10.0	1.0
Littleton Formation	Dbl	2	2	0	0	0.4
Manhattan Schist	Cm	5	5	0	0	1.8
Maromas Granite Gneiss	Dm	1	1	0	0	<0.2
Oronoque Schist	Oo	1	1	0	0	0.4
Ratlum Mountain Schist	Or	47	40	0	2.5	4.0
Rowe Schist	OCr	14	14	0	0	1.7
schist and granulite member of Trap Falls Formation	Otfg	17	14	0	0	0.5
Scotland Schist	DSs	3	3	0	0	0.9
Shelton (white gneiss) Member of Trap Falls Formation	Otfs	3	3	0	0	0.3
Southington Mountain Member of The Straits Schist	DSts	1	1	0	0	0.3
Taine Mountain Formation	Ot	9	9	0	0	0.8
Taine Mountain Formation and Collinsville Formation undivided	Ot + Oc	4	4	25.0	25.0	0.3
The Straits Schist	DSt	12	12	8.3	0	1.9
Walloomsac Schist	Ow	6	6	0	0	0.5
Wepawaug Schist	DSw	9	8	11.1	0	0.7
Whigville Member of Taine Mountain Formation	Otwv	1	1	0	0	<0.2
Sulfidic schists						
Brimfeld Schist	Obr	33	32	0	0	3.2
Carringtons Pond Member of Trap Falls Formation	Otfc	3	3	33.3	0	0.5
rusty mica schist and gneiss	Ygs	2	2	0	0	0.9

Geologic unit name	Geologic unit code	Number of samples		MCL, percentage1 of water samples with concentrations, in micrograms per liter		Percentage of study area underlain by geologic unit2
		Arsenic	Uranium	Arsenic >10 μg/L	Uranium >30 μg/L	
upper slice of Canaan Mountain Schist	Cmcu	1	1	0	0	0.3
unmapped areas	NA	1	1	0	0	1.0
Overall for the study area	NA	660	589	3.6	3.9	82.4

[Geologic unit names are the Connecticut Department of Energy and Environmental Protection preferred names as modified from Rodgers (1985). Bedrock categories (subheadings) are modified from Robinson and Kapo (2003). Color shadings indicate the percentage of wells with exceedances above concentration thresholds in ranges of , no data (—); , less than (<) 1 percent; , 1 to 10 percent; , more than (>) 10 to 20 percent; , >20 to 30 percent; and , >30 percent. MCL, U.S. Environmental Protection Agency maximum contaminant level enforceable for public water supplies; μg/L, microgram per liter; NA, not available]

[1]The percentage (of exceedance) values for geologic units computed for this study may not represent the actual risk for existing and future wells in these units.

[2]About 17.6 percent of the study area (State of Connecticut) was underlain by 75 individual geologic units from which no water samples were collected.

Table 4. Arsenic and uranium concentrations that exceed maximum contaminant levels from private wells in Connecticut, for three towns, 2013–15

Town	Number of samples		Percentage of samples with concentrations that exceed MCL	
	Arsenic	Uranium	Arsenic >10 μg/L	Uranium >30 μg/L
Stamford	732	731	0.1	11.2
Weston	110	—	47.3	—
Wilton	81	80	2.5	7.5
Statewide dataset	660	589	3.6	3.9

[MCL, U.S. Environmental Protection Agency maximum contaminant levels enforceable for public water supplies; μg/L, microgram per liter; >greater than; —, no data].

HUMAN HEALTH IMPLICATIONS

High concentrations of arsenic and uranium in drinking water have been associated with increased risk of lung, bladder, and skin cancers for arsenic and adverse effects on kidney function for uranium (Magdo and others, 2007). For households in Connecticut that rely on private wells for drinking water, this study has shown that the likelihood of having arsenic or uranium concentrations that exceed human-health benchmarks may depend in large part on the particular geologic unit in which the household's well is located. Other factors can affect arsenic or uranium concentrations, such as geochemical conditions or residence time for water-rock reactions in the local groundwater system.

References

Ayotte, J. D., Montgomery, D. L., Flanagan, S. M., and Robinson, K. W., 2003, Arsenic in groundwater in eastern New England: occurrence, controls, and human health implications: *Environmental Science and Technology*, v. 37, no. 10, p. 2075- 2083. [Also available at https://nh. water.usgs.gov/Publications/2003/es026211g.pdf].

Ayotte, J. D., Nolan, B. T., Nuckols, J. R., Cantor, K. P., Robinson, G. R., Jr., Baris, Dalsu, Hayes, Laura, Karagas, Margaret, Bress, William, Silverman, D. T., and Lubin, J. H., 2006, Modeling the probability of arsenic in groundwater in New England as a tool for exposure assessment: *Environmental Science and Technology*, v. 40, p. 3578–3585.

Colman, J. A., 2011, Arsenic and uranium in water from private wells completed in bedrock of east-central Massachusetts— Concentrations, correlations with bedrock units, and estimated probability maps: *U.S. Geological Survey Scientific Investigations Report* 2011–5013, 113 p.

Connecticut Department of Public Health, 2013, *Private well testing: Connecticut Department of Public Health* Publication no. 24, 6 p. [Also available at http://www.ct.gov/dph/lib/dph/environmental _health/eoha/pdf/24_residential_drinking_water_well_testing.pdf].

Connecticut Department of Public Health, 2016, *Private wells: Connecticut Department of Public Health Web page*, accessed March 29, 2016, at http://www.ct.gov/ dph/cwp/view.asp?a=3140&q=400544.

Flanagan, S. M, Belaval, Marcel, and Ayotte, J. D., 2014, Arsenic, iron, lead, manganese, and uranium concentrations in private bedrock wells in southeastern New Hampshire, 2012–2013: *U.S. Geological Survey Fact Sheet* 2014–3042, 6 p., appendixes. [Also available at https://doi. org/10.3133/fs20143042].

Flanagan, S.M., and Brown, C.J., 2017, Inventory of water-quality and geologic-setting data from 674 private wells in Connecticut, 2013-2015: U.S. *Geological Survey data release*, https://doi.org/10.5066/ F7K935P5.

Magdo, S. H., Forman, Joel, Graber, Nathan, Newman, Brooke, Klein, Kathryn, Satlin, Lisa, Amler, R. W., Winston, J. A., and Landrigan, P. J., 2007, Grand rounds— Nephrotoxicity in a young child exposed to uranium from contaminated well water: *Environmental Health Perspectives*, v. 115, no. 8, p. 1237–1241.

Montgomery, D. L., Ayotte, J. D., Carroll, P. R., and Hamlin, Patricia, 2003, Arsenic concentrations in private bedrock wells in southeastern New Hampshire: *U.S. Geological Survey Fact Sheet* 03–051, 6 p. [Also available at https://pubs.er. usgs.gov/publication/fs05103].

Robinson, G. R., Jr., and Ayotte, J. D., 2006, The influence of geology and land use on arsenic in stream sediments and ground waters in New England, USA: *Applied Geochemistry*, v. 21, p. 1482–1497.

Robinson, G. R., Jr., and Kapo, K. E., 2003, Generalized lithology and lithogeochemical character of near-surface bedrock in the New England region: *U.S. Geological Survey Open-File Report* 03–225, 60 p. [Also available at https://pubs.er.usgs.gov/publication/ofr03225].

Rodgers, John, comp., 1985, Bedrock geological map of Connecticut: Connecticut Geological and Natural History Survey, *Natural Resource Atlas Map Series*, scale 1:125,000, 2 sheets, accessed December 12, 2016, at https://ngmdb.usgs.gov/Prodesc/proddesc_54245.htm.

U.S. Environmental Protection Agency, 1994, *Methods for the determination of metals in environmental samples, supplement I* (rev. 5.4): U.S. Environmental Protection Agency EPA/600/R–94/111, 260 p.

U.S. Geological Survey, 2016, *The Branch of Quality Systems*: accessed November 2016, at https://bqs.usgs.gov.

In: Drinking Water
Editor: Erwin Jon Orosz

ISBN: 978-1-53617-284-3

Chapter 4

MANAGEMENT WEAKNESSES DELAYED RESPONSE TO FLINT WATER CRISIS*

Stacey Banks, Charles Brunton, Kathlene Butler, Allison Dutton, Tiffine Johnson-Davis, Fred Light, Jayne Lilienfeld-Jones, Tim Roach, Luke Stolz, Danielle Tesch and Khadija Walker

EPA Region 5 emergency response vehicle in Flint, Michigan. (EPA photo).

* This is an edited, reformatted and augmented version of U.S. Environmental Protection Agency Office of Inspector General Report No. 18-P-0221, dated July 19, 2018.

ABBREVIATIONS

CCT	Corrosion Control Treatment
CFR	Code of Federal Regulations
EPA	U.S. Environmental Protection Agency
FY	Fiscal Year
GAO	U.S. Government Accountability Office
LCR	Lead and Copper Rule
MDEQ	Michigan Department of Environmental Quality
OECA	Office of Enforcement and Compliance Assurance
OIG	Office of Inspector General
OW	Office of Water
ppb	parts per billion
PQL	Practical Quantitation Limit
PWSS	Public Water System Supervision
SDWA	Safe Drinking Water Act

AT A GLANCE

Why We Did This Project

The U.S. Environmental Protection Agency's (EPA's) Office of Inspector General (OIG) conducted this audit to examine the circumstances of, and the EPA's response to, contamination in the city of Flint, Michigan's community water system, including the EPA's exercise of its oversight authority. After Flint switched its drinking water supply in April 2014, inadequate water treatment exposed many Flint residents to lead.

In October 2016, the OIG issued a Management Alert to the EPA about specific factors that delayed its emergency response. Our current report evaluates additional matters concerning the agency's management controls when responding to the Flint contamination incident.

According to the MDEQ, more than $400 million in public funds has been spent by the EPA and Michigan to address the Flint crisis. Eight other federal agencies also provided support.

MANAGEMENT WEAKNESSES DELAYED RESPONSE TO FLINT WATER CRISIS

What We Found

The circumstances and response to Flint's drinking water contamination involved implementation and oversight lapses at the EPA, the state of Michigan, the Michigan Department of Environmental Quality (MDEQ), and the city of Flint. Since January 21, 2016, the EPA has overseen the implementation of its emergency administrative order and amendment issued in response to the drinking water contamination.

EPA Region 5 and EPA headquarters officials have worked with the MDEQ and Flint personnel to help improve the city's water system. As of May 2018, the state of Michigan and city of Flint have completed some actions and are working on remaining actions.

> The EPA should strengthen its oversight of state drinking water programs to improve the efficiency and effectiveness of the agency's response to drinking water contamination emergencies.

Michigan

Under the MDEQ's supervision, the Flint water system did not adhere to two Lead and Copper Rule requirements: (1) develop and maintain an inventory of lead service lines needed for sampling, and (2) maintain corrosion control treatment after the water source switch in April 2014. The rule requires utilities to minimize consumers' exposure to lead in drinking water. As the primacy agency, the MDEQ is responsible for enforcing this rule for Michigan water systems. The MDEQ did not issue a notice of violation or take other formal enforcement action regarding either requirement until August 2015. Instead, the MDEQ advised Flint public water system staff to conduct additional tests and to delay corrosion control treatment installation. The decision to delay corrosion control treatment prolonged residents' exposure to lead.

The EPA

The agency retains oversight and enforcement authorities to provide assurance that states with primacy comply with Safe Drinking Water Act requirements, such as those in the Lead and Copper Rule. However, Region 5 did not implement management controls that could have facilitated more informed and proactive decision-making when Flint and the MDEQ did not properly implement the Lead and Copper Rule. While Flint residents were being exposed to lead in drinking water, the federal response was delayed, in part, because the EPA did not establish clear roles and responsibilities, risk assessment procedures, effective communication and proactive oversight tools.

Recommendations and Planned Agency Corrective Actions

We recommend that EPA headquarters and EPA Region 5 use lessons learned from Flint to improve its oversight of Safe Drinking Water Act compliance. We also recommend that EPA headquarters revise the Lead and Copper Rule to improve the effectiveness of monitoring requirements. The EPA agreed with eight of the nine recommendations and provided adequate planned corrective actions and completion dates. Eight recommendations are resolved pending completion of corrective actions. Recommendation 1 is unresolved.

July 19, 2018
Memorandum

SUBJECT:	Management Weaknesses Delayed Response to Flint Water Crisis Report No. 18-P-0221
FROM:	Arthur A. Elkins Jr.
TO:	Susan Bodine, Assistant Administrator Office of Enforcement and Compliance Assurance David P. Ross, Assistant Administrator Office of Water Cathy Stepp, Regional Administrator Region 5

This is our report on the subject audit conducted by the Office of Inspector General (OIG) of the U.S. Environmental Protection Agency (EPA). The project number for this audit was OPE-FY16-0031. This report contains findings that describe the problems the OIG has identified and corrective actions the OIG recommends. This report represents the opinion of the OIG and does not necessarily represent the final EPA position. Final determinations on matters in this report will be made by EPA managers in accordance with established audit resolution procedures.

The EPA offices responsible for responding to issues in this report are the Office of Enforcement and Compliance Assurance; the Office of Water; and the Office of the Regional Administrator, Region 5.

Action Required

This report contains one unresolved recommendation. In accordance with EPA Manual 2750, the resolution process begins immediately with the issuance of this report. We are requesting a meeting within 30 days between the Assistant Administrator for Water and the OIG's Assistant Inspector General for Audit and Evaluation. If resolution is still not reached, the Office of Water is required to complete and submit a dispute resolution request to the Chief Financial Officer to continue resolution.

We will post this report to our website at www.epa.gov/oig.

INTRODUCTION

Purpose

The U.S. Environmental Protection Agency's (EPA's) Office of Inspector General (OIG) conducted this review to examine the circumstances of, and the EPA's response to, the contamination in the city of Flint, Michigan's community water system,[1] including the EPA's exercise of its oversight authority.

EPA OIG Report No. 17-P-0004, *Management Alert: Drinking Water Contamination in Flint, Michigan, Demonstrates a Need to Clarify EPA Authority to Issue Emergency Orders to Protect the Public*, issued October 20, 2016, alerted the EPA about specific factors that delayed its response. Our current report evaluates management controls at EPA Region 5 (responsible for Michigan), and EPA headquarters' response to the Flint contamination incident.

Background

Drinking Water Contamination in Flint, Michigan

In April 2014, Flint's water system, serving drinking water to a population of nearly 100,000 residents, switched from purchasing treated water from the Detroit Water and Sewerage Department (now called Great Lakes Water Authority) to sourcing and treating its own water supply from the Flint River.

Governor-appointed emergency managers charged with improving Flint's finances initially opposed this change, but ultimately implemented it as a cost-saving measure.

Since 1967, the Flint drinking water plant had only distributed drinking water purchased from Detroit. Flint emergency managers considered the use of the Flint River

[1] The Safe Drinking Water Act and its regulations classifies Flint's public water system as a large community drinking water system, hereafter referred to as the Flint "water system."

to be temporary. They planned to purchase water from a new utility, the Karegnondi Water Authority, starting in 2016.

Flint Water Treatment Plant sign. (EPA OIG photo).

Detroit's water system included an additive that coated pipes to prevent corrosion. The city of Flint's process for treating water from the new source did not include treatment for reducing corrosion. Without this treatment, lead from components within the distribution system can leach into drinking water.

After the April 2014 source switch, consumers began complaining about water quality issues. In August 2014, EPA Safe Drinking Water Act (SDWA) data showed the city of Flint violated regulations for total coliform (bacteria), and in December 2014, total trihalomethanes (a byproduct of the disinfection process). In February 2015, citizens started reporting concerns about lead contamination to EPA Region 5.

Appendix A contains a detailed timeline of events associated with the Flint drinking water incident.

SDWA and State Primacy

The SDWA, 42 U.S.C. § 300f et seq., is the federal law that protects public drinking water supplies throughout the nation. The law assigns the EPA Administrator the ultimate authority to protect public health by setting and enforcing drinking water quality standards. However, the SDWA allows the EPA to grant states the authority to implement and enforce SDWA regulations in an arrangement referred to as "primacy."

To receive EPA approval for primacy, states must demonstrate that they have adopted regulations that are at least as stringent as national requirements, are implementing adequate procedures for the enforcement of these regulations, and will keep records and make reports as the EPA may require. The EPA has granted nearly all states—including Michigan—primacy to implement the SDWA.[2] When states are granted primacy, the EPA retains the responsibility for overseeing state implementation and federal enforcement authority. In 1978, the EPA determined that Michigan met all SDWA requirements to be granted primacy, including adopting and implementing enforcement procedures.

Guidance from the EPA's Office of Enforcement and Compliance Assurance (OECA) instructs EPA regional offices to "ensure that primacy agencies fulfill the enforcement conditions of their primacy agreements."[3] The EPA uses a variety of tools to oversee state drinking water implementation, such as providing technical assistance and regularly conferring with state agencies. Specifically, the SDWA authorizes the EPA to request information, take independent enforcement actions, and revoke state primacy when states do not implement the SDWA with the stringency required by federal law.

Sign directing residents to water distribution center during response efforts (EPA photo).

[2] The state of Wyoming, Washington, D.C., and most tribes are not authorized to implement the SDWA.
[3] OECA National Program Manager Guidance for fiscal years (FYs) 2016–2017.

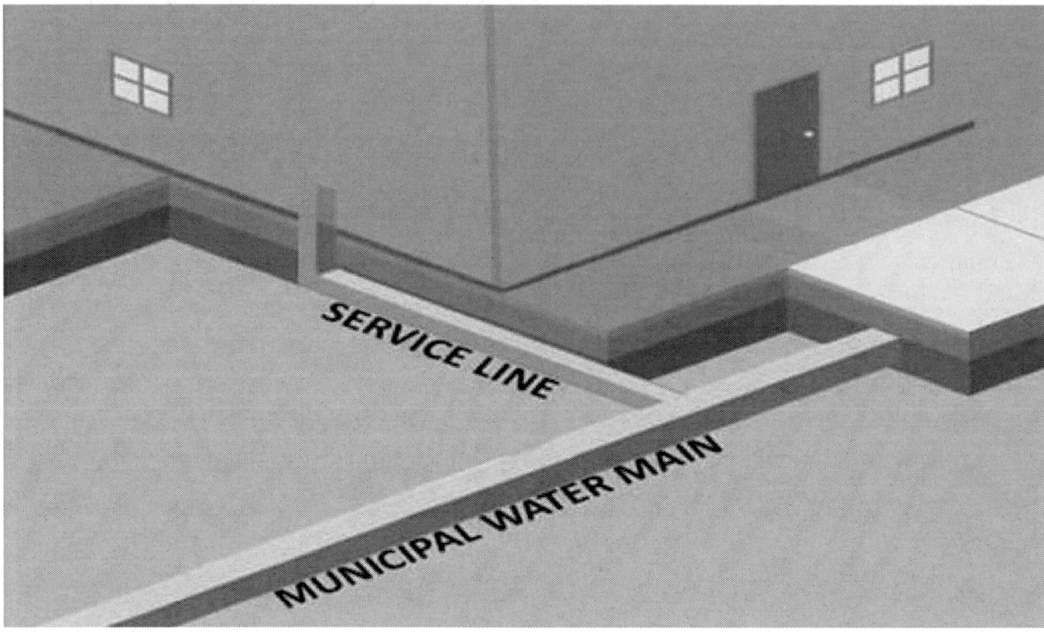

Source: An EPA OIG image.

Figure 1. Diagram of a distribution system showing a service line and water main.

Lead Contamination and SDWA

Drinking water distribution systems include water main pipes and service line pipes (Figure 1). These systems may contain lead or copper components.

Reduce Your Exposure to Lead in Drinking Water at Home

- Use only cold water for drinking, cooking and making baby formula. Boiling water does not remove lead from water.
- Regularly clean your faucet's screen (also known as an aerator).
- Consider using a water filter certified to remove lead and know when it is time to replace the filter.
- Before drinking, flush your pipes by running your tap, taking a shower, doing laundry or a load of dishes.

Source: The EPA's *Basic Information About Lead in Drinking Water* webpage.

Drinking water leaving a treatment plant may be generally free of lead, but the water can corrode pipes in the distribution system and plumbing materials in homes, and release lead into drinking water. For additional information, see the EPA's "Sources of Lead in Drinking Water" factsheet.

In children, exposure to lead can cause serious health problems, including the potential for developmental disorders, lower IQs and delinquent behavior. In adults, lead exposure may result in poor pregnancy outcomes, dementia and cardiovascular disease.

In 1991, the EPA issued the Lead and Copper Rule (LCR), 40 CFR § 141.80 et seq., to minimize lead and copper in drinking water.[4] Lead and copper are only two of over 90 contaminants regulated under the SDWA. The National Primary Drinking Water Regulations establish either a maximum contaminant level (a health-based standard), or a treatment technique requirement.

There is no safe level of lead in drinking water. However, the LCR established the treatment technique requirements to minimize exposure. These requirements compel drinking water systems to conduct tap sampling for lead and copper to determine the treatment techniques and other steps systems must take to reduce exposure. To adhere to the LCR, utilities must conduct monitoring for lead and copper in water systems, and demonstrate that they comply with monitoring and treatment technique requirements.[5] Figure 2 illustrates key components of the LCR relevant to the city of Flint.

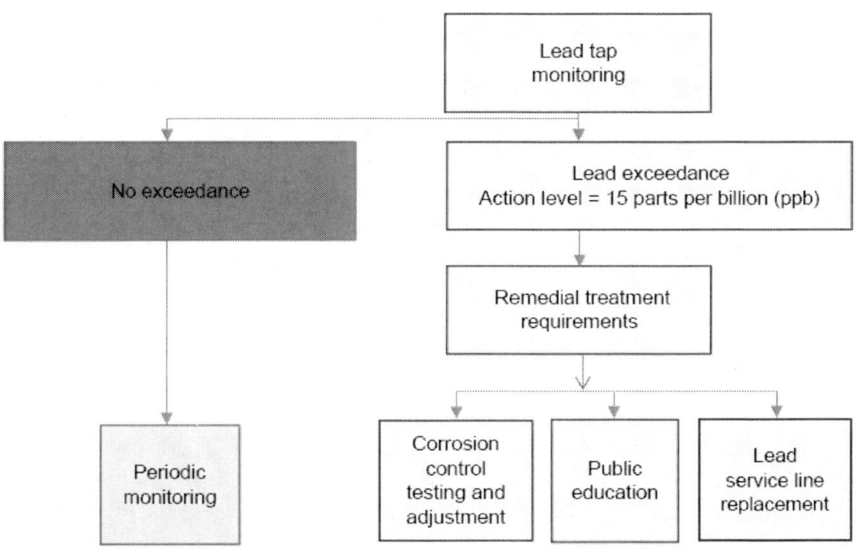

Source: An EPA diagram modified by the OIG. The red outline boxes are discussed in sections below.

Figure 2. Key components of the LCR relevant to Flint.

---

[4] Some LCR requirements vary based on the water system size and type. Since Flint is a large community system, this report discusses the requirements for large community systems—those serving more than 50,000.

[5] In 1986, Congress amended the SDWA to prohibit the use of pipes, solder or flux that are not "lead free" in public water systems, or in plumbing where facilities provide water for human consumption. In 1996, Congress further amended the SDWA by prohibiting the sale of any pipe or plumbing fixture that was not lead free, except for a pipe used in manufacturing or industrial processing. In 2000, the EPA published revisions to the LCR to address implementation issues arising from legal challenges to the 1991 rule. The revisions also streamlined and reduced the monitoring and reporting burden. In 2007, the EPA revised the LCR to enhance implementation in the areas of monitoring, treatment, customer awareness and lead service line replacement.

Treatment Technique Requirements: Corrosion Control (40 CFR § 141.81 and § 141.82)

The LCR requires community water systems to optimize corrosion control. Utilities have optimized corrosion control when the treatment method that they use has minimized the potential for corrosion in the distribution system, thus minimizing lead contamination. The rule does not prescribe specific corrosion control treatments. Instead, there is a range of methods that water systems can use.

When the LCR was implemented in 1991, it established timelines for the installation of corrosion control treatment and for subsequent monitoring.

The LCR required large water systems to install optimal corrosion control treatment by January 1, 1997, unless they qualified for certain exemptions.[6]

LCR Terms

Action level of 15 ppb: Utilities are required to monitor so that lead concentrations in the 90th percentile do not exceed the *federal action level* of 15 ppb.

Water quality parameters: State-approved measures that are used to determine whether corrosion control is working within the distribution system.

Practical Quantitation Limit (PQL) of 5 ppb: Level of lead concentration in a large system where the LCR required optimal corrosion control treatment.

One exemption was to demonstrate low lead levels at consumers' taps, where water systems monitor drinking water over two consecutive 6-month periods. Water systems were required to show whether lead concentrations fell at or below 5 ppb (the PQL) between the entrance to the distribution system and consumers' taps. If results exceeded 5 ppb, water systems were required to establish and optimize corrosion control treatment.

According to the LCR, if a water system has optimized its corrosion control treatment and plans to change water sources or drinking water treatment methods, the state must review and approve plans before the change can be implemented.

The LCR requires states to review and approve the optimized corrosion control treatment for all systems. The LCR also requires states to designate optimal water quality parameters intended to represent the conditions under which systems must operate their corrosion control treatment. The rule requires that water systems measure against those parameters, which are typically set as ranges or minimums. Without set water quality parameters, a state does not have a reliable method for gauging the adequacy of corrosion control treatment or determining compliance.

The LCR allows the EPA to intervene if a state fails to approve appropriate treatment plans. The rule authorizes EPA regional administrators to review state treatment determinations and issue alternate federal determinations, if the EPA finds the following:

[6] The exemption criteria are specified in 40 CFR § 141.81(b)(2)-(b)(3).

Drinking water samples taken from Flint residences during the emergency response. (EPA OIG photo, taken April 2016).

- A state has failed to issue a treatment determination by the applicable deadlines,
- A state has abused its discretion in a substantial number of cases or in cases affecting a substantial population, or
- The technical aspects of a state's determination would be indefensible in an expected federal enforcement action taken against a system.[7]

Tap Water Monitoring for Lead (40 CFR § 141.86)

After a system has optimized corrosion control, the LCR requires water systems to monitor for changes in lead concentration on a regular basis. These tests verify that when drinking water reaches the consumer, lead concentrations do not exceed the federal action level of 15 ppb in more than 10 percent of homes.[8]

Lead service lines showing inner surface without any coating from corrosion control treatment (left), with coating (right), and fully corroded (middle) (EPA photo).

[7] 40 CFR § 142.19 provides the procedures for the EPA Administrator's review of state determinations and issuance of a federal order.

[8] 40 CFR § 141.80(c)(1) describes this as the "90th percentile" lead level exceeding 0.015 milligrams per liter.

To conduct the monitoring, the LCR requires utilities to identify the highest-priority sampling sites, such as single-family homes served by lead service lines. The highest-priority sampling sites are identified as "tier 1" locations and, for Flint, constitute the entire sampling pool for regular testing.

The LCR instructs water systems to regularly report tap water sample results from tier 1 sites. Water systems monitor tap water on a semiannual, annual or triennial basis to verify lead contamination does not exceed the federal action level. When a large water system exceeds that level in more than 10 percent of home tap water samples, the water system must provide public education on lead, remove a percentage of lead service lines, and conduct source water monitoring.

Public Education (40 CFR § 141.85)

The LCR requires systemwide public education when tap-monitoring results show that a system exceeds the federal action level for lead. There are also supplemental lead-monitoring requirements for all water systems to provide consumer notices to those whose taps were tested. The purpose of public education is to inform consumers about lead results, the health effects of lead, and steps the public can take to reduce exposure.

Water systems must submit all written public education materials to the state prior to releasing the information to the public.

Lead Service Line Replacement (40 CFR § 141.84)

In the event of a lead action level exceedance, the LCR requires water systems to identify the number of lead service lines present in the distribution system, and requires water systems to meet the minimum 7 percent replacement rate through either physically replacing lead service lines or individually testing them to demonstrate their lead concentrations are below 15 ppb. However, if the system returns to a lead level below the action level after two monitoring periods, the utility is not required to continue removing lead service lines. In addition, if a system is in violation of 40 CFR § 141.81 for failure to install corrosion control treatment, the state can require the system to commence lead service line replacement until the system is below the lead action level for two consecutive monitoring periods.

Emergency Administrative Order and Amendment

On January 21, 2016, OECA issued an Emergency Administrative Order requiring the state of Michigan, the Michigan Department of Environmental Quality (MDEQ), and the city of Flint to take steps to reduce lead contamination.

A bottled water distribution site in Flint. (EPA OIG photo, taken April 2016).

EPA managers monitoring compliance with the order determined that while the state of Michigan, the MDEQ, and the city of Flint made a good-faith effort to comply, the EPA did not see progress in completing some of the required actions. In November 2016, EPA Region 5 issued its First Amendment to Emergency Administrative Order. The amendment set new goals and allowed the city to develop milestones it could achieve.

Subsequently, OECA and EPA Region 5 have overseen the respondents' compliance with both the order and the amendment to the order (Appendix B). According to Region 5, the EPA worked with the city and the MDEQ to do the following:

- Oversee treatment plant operations.
- Provide outreach to impacted residents.
- Mobilize the largest sampling effort in EPA history.
- Address technical problems through a cross-agency expert task force.
- Hold open-house information sessions for residents.
- Hold science summits with federal, state and local scientists and experts.

The EPA maintains a list of activities and progress updates for the order and its amendment on the agency's Flint Drinking Water Documents webpage. The MDEQ reported that by December 2017, lead levels in Flint's drinking water had fallen below the federal action level of 15 ppb.

However, the EPA and local authorities continued to recommend that residents filter their water for drinking and cooking. As of May 2018, the state of Michigan and the city of Flint have not completed all actions in the order and amendment.

Please see Appendix B for status of order and amendment actions. Currently, the city and state are working to replace lead service lines.

A billboard in the city of Flint. (EPA OIG photo, taken April 2016).

Through its Public Water System Supervision (PWSS) program, the EPA awarded Michigan an average of $4.2 million annually to fund over 11,000 water systems in the state.[9] According to the EPA, as of June 2016, the agency spent more than half of that amount (over $3 million) on the Flint drinking water contamination incident in the first 6 months after the order was issued (Figure 3). The state of Michigan provided approximately $250 million to address problems in Flint's water system, and to increase the city's technical and managerial capacity.

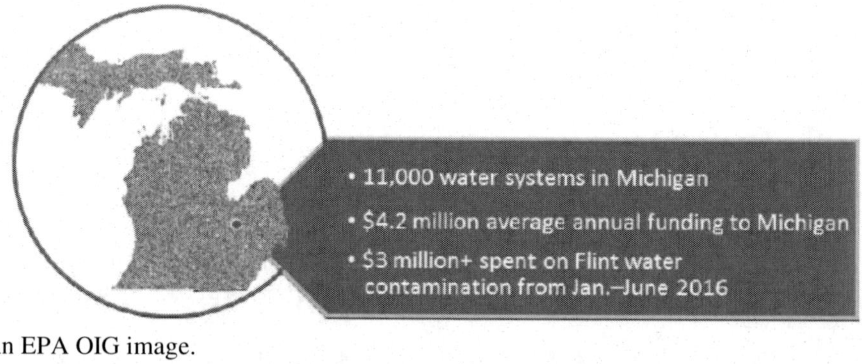

Source: An EPA OIG image.

Figure 3. Comparison of Michigan's statewide water system funding and Flint contamination spending.

In March 2017, the EPA awarded an additional $100 million[10] to the MDEQ for Flint water infrastructure upgrades, which required a state match of 20 percent or $20 million. In September 2017, the MDEQ reported that combined federal and state spending to address the contamination totaled more than $400 million.[11]

[9] Average of funding awards in FYs 2011–2016.
[10] EPA press release: *EPA Awards $100 Million to Michigan for Flint Water Infrastructure Upgrades* (March 17, 2017).
[11] In addition to the EPA, other federal agencies also provided support and funding.

Challenges with the LCR

Assuring public health under the LCR presents challenges. A key challenge is that the LCR is based on treatment, not health-based requirements. The EPA has recognized the need to clarify and strengthen the rule, and the agency is in the process of revising it.

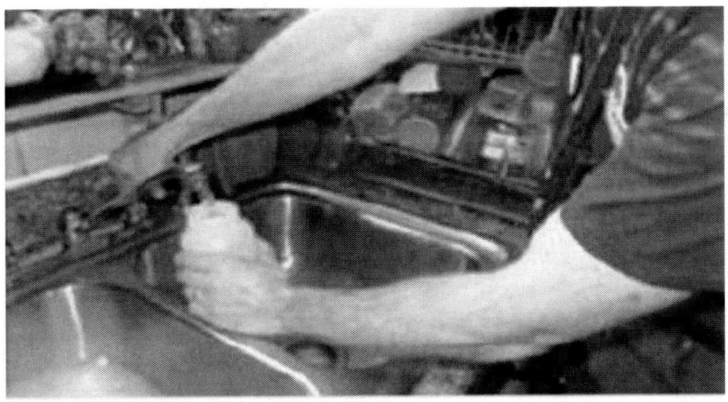

An EPA employee collects a water sample to send to a laboratory for analysis during response efforts. (EPA photo).

In October 2016, the EPA's Office of Water released the *Lead and Copper Rule Revisions White Paper* for the purpose of keeping the public informed about options being considered by the agency. The report suggests that proposed revisions to the rule should include technology-driven and health-based elements to avoid high lead levels and health risks. Some of the elements under consideration include the following:

- Incorporating health-based goals for lead to guide actions and communicate risk at the household level.
- Implementing and enforcing the mechanism for lead service line removal programs.
- Requiring all systems to implement corrosion control treatment, regardless of system size, tap sampling results, or presence of lead service lines.
- Using "point of use" filters when addressing risks from lead and copper at the household level.
- Clarifying and strengthening sampling requirements, reducing uncertainty, and ensuring broader and more consistent proactive application of corrosion control treatment.
- Increasing the transparency associated with implementing the LCR by publicly posting the locations of lead service lines and sampling results.
- Strengthening public education requirements in the LCR.

Management Controls for Government Programs

All federal agencies must establish certain standards, policies and practices to create a sound internal control structure for their programs. Internal controls help to prevent and detect errors, fraud and noncompliance with provisions of laws and regulations. Agencies and individual federal managers are required to develop internal controls for programs to achieve results and comply with applicable laws and regulations.

The U.S. Government Accountability Office (GAO) developed internal control standards in 1999 (revised in 2014) to assist agencies in achieving their objectives. The standards focus on effective internal control through communication, risk assessment and oversight. The EPA instructed managers to use these standards in the agency's EPA Order, *Management's Responsibility for Internal Control* (2008).

Likewise, the EPA's FY 2017 Guidance for Enterprise Risk-Based Decision Making at EPA: Integrating Strategic Reviews and Management Integrity Internal Controls incorporates GAO guidelines and tools to integrate risk assessment into the strategic planning process.

Responsible Offices

EPA Region 5 drinking water program staff and managers primarily oversee SDWA implementation in regional primacy states, including Michigan. Two headquarters offices help EPA regions implement the SDWA:

- The Office of Water (OW) provides programmatic and implementation guidance.
- The Office of Enforcement and Compliance Assurance (OECA) advises EPA regions about enforcing SDWA provisions.

Prior Reports

EPA OIG Report No. 17-P-0004, titled *Management Alert: Drinking Water Contamination in Flint, Michigan, Demonstrates a Need to Clarify EPA Authority to Issue Emergency Orders to Protect the Public*, was issued October 20, 2016. The report noted that the EPA needs to clarify for its employees how its emergency authority can and should be used to intervene in a public health threat. The OIG recommended that OECA update the EPA's 1991 guidance on SDWA § 1431 emergency authority. We also recommended that OECA require all relevant EPA drinking water, and water enforcement program management and staff, attend training on SDWA § 1431 authority. The agency agreed with the recommendations and corrective actions are in progress.

GAO Report No. GAO-17-424, titled *Drinking Water: Additional Data and Statistical Analysis May Enhance EPA's Oversight of the Lead and Copper Rule*, is a congressionally requested report dated September 2017. The report found that available EPA data reported by states show that of the approximately 68,000 water systems subject to the LCR, at least 10 percent had at least one open violation of the rule. However, these and other data are not complete. GAO made three recommendations, including one recommendation that the EPA require states to report data on lead pipes, and develop a statistical analysis on the likelihood of LCR violations to supplement the EPA's current oversight. The EPA agreed with the GAO's recommendations.

Scope and Methodology

We conducted this performance audit from February 2016 through April 2018, in accordance with generally accepted government auditing standards. Those standards require that we plan and perform the audit to obtain sufficient, appropriate evidence to provide a reasonable basis for our findings and conclusions based on our audit objectives. We believe that the evidence obtained provides a reasonable basis for our findings and conclusions based on our audit objectives.

We reviewed the laws, regulations, policies, procedures and guidance related to the SDWA program. At EPA headquarters, we interviewed the former EPA Administrator; the former EPA Deputy Administrator; and staff and officials from the EPA's Office of General Counsel, the Office of Water, and the Office of Enforcement and Compliance Assurance. We also interviewed EPA Region 5 staff and officials, including the former Regional Administrator (who served until January 2016), and the former acting Regional Administrator (who served from January 2016 through January 2018). Further, we interviewed MDEQ Office of Drinking Water and Municipal Assistance personnel, former and current employees of the city of Flint, residents of Flint, and external experts. We also reviewed documents from the EPA and the MDEQ.

Circumstances Leading to Flint Drinking Water Contamination and EPA Emergency Administrative Order

Under the MDEQ's oversight, the Flint water treatment plant did not adhere to two LCR requirements: (1) identify and maintain a pool of tier 1 sampling sites, and (2) install and maintain corrosion control treatment throughout the system. State and local decisions to not adhere to these LCR requirements led to the corrosion of the Flint distribution system, which exposed residents to lead in their drinking water.

Inventory of Required Tier 1 Locations Was Not Maintained

The city of Flint did not develop or maintain accurate records of lead service line locations to identify tier 1 sampling sites. Primacy states like Michigan must require water systems to collect and maintain lead service line information, in accordance with the LCR.[12] Without this information, Flint could not prioritize its sampling efforts to collect water samples where higher levels of lead contamination were most likely to occur, as required by the LCR.

The January 2016 EPA Emergency Administrative Order required Flint to provide the agency with the city's existing inventory of homes with lead service lines.

When the city submitted the inventory, the EPA found that it was not complete or accurate. However, Flint did not update its inventory of lead service lines until August 2016. EPA Region 5 staff stated that other cities face similar challenges with inventories for lead service lines in their water systems. The EPA identified this as a nationwide weakness in its October 2016 *Lead and Copper Rule Revisions White Paper*.

Corrosion Control Treatment Was Not Maintained

The MDEQ did not require Flint's water system to maintain corrosion control treatment when Flint changed water sources in April 2014. According to EPA Region 5, the MDEQ concluded that Flint's change in source water would require Flint to revert to the LCR provision that required the water system to conduct tests to determine whether corrosion control treatment was necessary.[13] To do this, MDEQ personnel instructed Flint system staff to conduct monitoring during two consecutive 6-month periods to determine whether corrosion control treatment was necessary.

EPA Region 5 disagreed with the MDEQ's interpretation. The LCR treats systems (such as Flint's) that change their source water or treatment as an existing system. As a result, Region 5 concluded that Flint's system was an existing system with a new source. Therefore, under the LCR, the city needed to maintain continuous corrosion control treatment.

In January 2015, results from Flint's first round of testing found a 90th percentile lead concentration of 6 ppb. However, the MDEQ did not instruct Flint to install corrosion control treatment (per the LCR) but instead required an additional 6-month round of sampling. In July 2015, results from the second round of 6-month tests showed 90th percentile lead concentrations had increased to 11 ppb.

At that time, EPA Region 5 and the MDEQ reached an agreement to require the Flint water system to begin corrosion control treatment as soon as possible. In August 2015,

---

[12] 40 CFR § 141.86(1); 40 CFR § 141.91.

[13] 40 CFR § 141.81(b)(3). EPA Region 5 learned about the MDEQ's conclusion in April 2015.

the MDEQ required Flint to select a corrosion control treatment (no later than January 2016).

Excavated lead service lines used to test water chemistry at the Flint water treatment plant (EPA OIG photo).

In August and September 2015, a private researcher conducted testing and identified the following:

- 110 homes sampled with lead levels over 5 ppb.
- 43 homes sampled with lead levels over 15 ppb.
- 7 homes sampled with lead levels over 100 ppb.
- 1 home sampled with a lead level over 1,000 ppb.

In September 2015, a local doctor issued a study showing an increase in the blood lead levels of children living in Flint.

In October 2015, instead of installing corrosion control treatment, Flint's water system returned to purchasing drinking water from the Great Lakes Water Authority (formerly called the Detroit Water and Sewerage Department), which already included corrosion control treatment. At the time of this change, almost a year-and-a-half of exposure to improperly treated water had damaged the city's drinking water infrastructure, and lead concentrations continued to rise. In December 2015, the MDEQ reported that Flint began supplementing Detroit water with additional corrosion control treatment. However, due to the damage done to the Flint distribution system, the lead levels in drinking water did not fall below the federal action level until late 2016.

Conclusion

The MDEQ did not enforce LCR provisions that require Flint's water system to keep an accurate and complete inventory of tier 1 sample site locations, or maintain corrosion control treatment. As a primacy agency, the MDEQ bore responsibility for advising Flint's water system staff about the drinking water source change and meeting SDWA standards. However, MDEQ personnel misinterpreted the law, which led them to provide incorrect advice to Flint on corrosion control treatment requirements. This resulted in infrastructure damage and the prolonged exposure of Flint residents to lead in their drinking water.

Recommendations

We recommend that the Assistant Administrator for Enforcement and Compliance Assurance, and the Assistant Administrator for Water:

- Establish controls to annually verify that states are monitoring compliance with all Lead and Copper Rule requirements, including accurately identifying tier 1 sampling sites and maintaining continuous corrosion control treatment.

We recommend that the Assistant Administrator for Water:

- Revise the Lead and Copper Rule to improve the effectiveness of monitoring and corrosion control treatment protocols.

Agency Response and OIG Evaluation

The agency provided a coordinated response from EPA Region 5, the OW and OECA. The offices provided planned corrective actions in their response, and additional clarifications during a June 5, 2018, meeting with the OIG.

OECA and the OW agreed with Recommendation 1; however, the offices did not provide acceptable corrective actions to address the recommendation. In their original response, the offices stated that they would develop metrics related to LCR implementation and incorporate the metrics into the regional protocol for conducting annual PWSS program reviews by September 30, 2019. The two offices also would implement a national oversight approach for drinking water programs by June 30, 2019.

The EPA provided supplemental information in the June 5, 2018, meeting with the OIG. In that meeting, and in subsequent correspondence on June 13, 2018, OECA and the OW revised their response to state that they agree on the value of national oversight for compliance monitoring and enforcement of the EPA's Drinking Water Program, but they said it is vital that they evaluate and develop it in a collaborative manner with their state partners. The offices planned to work with states to develop an approach or a pilot for implementing a national compliance monitoring and enforcement oversight program for drinking water, as appropriate, by June 2019.

The OIG does not agree that developing a pilot or approach constitutes establishing annual controls, per the recommendation language. The OIG did not accept this response as meeting the intent of Recommendation 1, and this recommendation remains unresolved.

The OW agreed with Recommendation 2. To address this recommendation, the OW will continue to evaluate input received from state, local and tribal partners, and evaluate the best available peer-reviewed science to ensure that the revised rule reflects the best ways to improve public health protection. The OW plans to complete this action by February 28, 2019. Based on the response and clarifications, the EPA provided acceptable planned corrective actions and completion dates for Recommendation 2.

Region 5, OECA and the OW provided supplemental technical comments that the OIG incorporated as appropriate. The EPA's coordinated response is found in Appendix C.

EPA Region 5's Management Weaknesses Delayed Federal Intervention

The EPA retains oversight and enforcement authorities to provide assurance that primacy states comply with the SDWA. However, timely oversight interventions rely on effective management systems that govern how and when the agency should intervene. EPA Region 5 did not manage its drinking water oversight program in a way that facilitated effective oversight and timely intervention in Flint. EPA Region 5 did not:

- Establish clear roles and responsibilities with the MDEQ.
- Communicate clearly and effectively.
- Use effective risk assessment protocols.
- Proactively use available SDWA authorities and oversight tools to intervene in Michigan's drinking water program.

These weaknesses limited Region 5's ability to monitor, adapt and respond to changing situations in Michigan and the city of Flint.

Region 5 and MDEQ Did Not Establish Clear Oversight Roles and Responsibilities

The EPA retains the authority and responsibility to oversee states with primacy over their drinking water programs. The EPA is empowered and required to intervene when states do not fulfill their responsibilities. However, EPA Region 5 staff and managers did not establish clear roles and responsibilities needed to foster a constructive federal-state relationship with the MDEQ's drinking water program staff. As previously mentioned, state primacy agreements obligate states to implement the SDWA in a manner that is at least as stringent as federal requirements. Further, the EPA's National Program Manager Guidance directs EPA regions to "ensure that primacy agencies fulfill the enforcement conditions of their primacy agreements" under the SDWA.

EPA Region 5 aimed to foster a collaborative relationship with the MDEQ.

The region's focus on maintaining a partnership influenced how Region 5 staff conducted oversight and enforcement in the state. In the case of the MDEQ, this partnership limited effective EPA oversight. While Region 5 needs to work with states to accomplish the mission of the agency, the region also needs to establish clear roles and responsibilities so that both the MDEQ and the region know how and when the EPA may intervene to implement and enforce the SDWA.

As a result of this relationship, Region 5 managers did not use their knowledge about the MDEQ's incomplete implementation of the SDWA when assessing the risks in Flint. For example, in 2010, an EPA contractor reviewed the MDEQ's drinking water program and identified MDEQ implementation deficiencies.[14] The review noted that the MDEQ disinvested from 10 SDWA requirements, which Michigan designated as temporary and non-health related.[15] However, the review concluded that the MDEQ should reconsider these disinvestments because at least one of them could impact public health. The contractor recommended that the MDEQ implement all of the SDWA requirements, as noted in the following quote:

MDEQ should reconsider the disinvestment activities. MDEQ's actions and policies should be as stringent as federally mandated rules and policies. All instances where the federal rules were not correctly implemented were treated as discrepancies.

In the 2010 report, Region 5 responded that the disinvestments were intended to be temporary and only include nonpublic-health-related primacy activities.

Region 5 said these disinvestments were allowed because the region viewed these disinvestments as an "innovative way for states to continue to ensure safe drinking water is provided to the public, even as states are struggling with diminishing resources."

---

[14] U. S. Environmental Protection Agency Public Water System Supervision Program, Final Report: Program Review for the Michigan Department of Environmental Quality Water Bureau, August 30, 2010.

[15] A disinvestment is a means to temporarily forfeit oversight functions required to implement the SDWA.
Due to limited resources, the MDEQ proposed, and EPA Region 5 approved, disinvestments on an annual basis.

Despite the recommendation from the contractor, these disinvestments continued and increased to 11 by 2015. Region 5 managers told us that being informed about implementation deficiencies in advance was a "proactive" approach to oversight. They maintained that disinvestments were more administrative and generally did not have a direct public health impact.

Table 1. Examples of MDEQ disinvestments and potential effects

SDWA requirement	Disinvestment	Potential effect
All water systems must report to the state the results of all tap samples on a defined schedule.	MDEQ would no longer issue violations for late reporting of monitoring results.	By not issuing these violations, identification of contamination could be delayed, further prolonging consumer exposure to contamination.
All systems must submit reporting forms to the state, including the location of each site and the criteria under which the site was selected for the system's sampling pool. Include an explanation of why sampling sites have changed.	MDEQ would no longer issue violations for failure to submit the LCR reporting form.	Without the LCR reporting forms, the state could not review information showing whether lead sampling met LCR requirements.
The state must require utilities to publish annual Consumer Confidence Reports with specific content requirements on a defined schedule (reports are public).	MDEQ would no longer issue violations for late reporting or insufficient content in Consumer Confidence Reports.	By not enforcing rules for Consumer Confidence Reports, the state loses a method for providing water system users complete and timely reports about their water quality.

Source: OIG analysis.

The MDEQ continued these disinvestments through 2015, and the OIG concluded that the disinvestments did have potential public health effects. However, Region 5 did not intervene to ensure that the MDEQ's drinking water program met minimum federal standards. It was not until 2016, while under the emergency order, that the MDEQ discontinued the majority of the disinvestments.[16]

In 2017, Region 5 began requiring states to submit implementation plans to describe how and when they would begin implementation of any primacy activities the states

[16] In 2016, an EPA manager told us that the MDEQ continued to be disinvested in three areas noted in the 2010 review.

could not fully implement at that time. Table 1 contains examples of MDEQ disinvestments and OIG analysis of potential effects.

EPA and MDEQ Did Not Communicate Effectively

Communication weaknesses contributed to a delayed federal response in Flint. For effective oversight, management needs accurate and complete information, and clear communication. However, the communication between the EPA and the MDEQ did not convey key information about human health risks from lead contamination in Flint. Communication within the EPA was also problematic.

These issues limited the EPA's knowledge about risks and contributed to the delayed federal response. For example:

- MDEQ personnel did not provide Region 5 with accurate information regarding corrosion control treatment. In February 2015, when EPA Region 5 staff asked about corrosion control, MDEQ personnel told them that Flint had an optimized corrosion control program in place, that the city conducted quarterly water quality parameter monitoring and did not have any unusual results, and that Flint continued to meet all applicable plant tap standards and treatment technique requirements. However, in April 2015, the state admitted that Flint was not using corrosion control treatment and argued that it was not required. Before that admission, Region 5 personnel did not have key information needed to intervene.
- Region 5 told us that communication with the MDEQ was frequent, consistent and clear, but the MDEQ failed to take appropriate actions when Region 5 consistently identified problems. According to EPA Region 5 managers, in April 2015, the EPA voiced concerns about the lack of corrosion control treatment to the MDEQ's Water Director. Region 5 managers said that in June 2015 they advised the MDEQ that the LCR required Flint to maintain consistent corrosion control treatment.

  However, an MDEQ manager stated that Region 5 did not advise them to initiate corrosion control at that point.
- Region 5 stated that the length of time that it took to obtain a legal interpretation delayed formal intervention. In a July 2015 meeting, the EPA and the MDEQ disagreed over interpretations of LCR corrosion control treatment requirements in Flint. The MDEQ requested, and Region 5 program staff agreed to obtain, a legal opinion from EPA headquarters. In August 2015, Region 5 and EPA headquarters began discussing Flint compliance with the requirements and the MDEQ's interpretation of those requirements.

However, headquarters personnel stated that Region 5's characterization of the situation lacked a sense of urgency. Region 5 did not make an official request for a headquarters' opinion until September 30, 2015.

Region 5 did not receive a legal opinion, but the region's request ultimately resulted in a November 2015 Office of Water memo clarifying the LCR corrosion control treatment requirements for all large water systems.

Although the memo mentioned Flint, it did not specifically address the Flint situation (see green box). For this reason, it was of limited use in dealing with the incident in Flint.

Summary of the "Lead and Copper Rule Requirements for Optimal Corrosion Control Treatment for Large Drinking Water Systems"

In this November 2015 memo, the EPA's Office of Water acknowledged differing interpretations of the LCR. The EPA's intent was to clarify that the rule requires any large system to continuously maintain corrosion control treatment.

A Region 5 official expressed concern that states would not follow guidance from the memo because it lacks the authority of law. In an effort to further clarify requirements, an EPA official stated that headquarters has provided training to EPA staff with regard to the memo.

Region 5 Did Not Have an Effective System for Risk Assessment

EPA staff reviewing lead sampling data in Flint during emergency response efforts. (EPA photo).

EPA Region 5 did not have effective risk assessment processes that would have alerted the region to growing issues in Flint. An effective risk assessment system provides the basis for responding to threats that impact human health. While bacterial violations alone would not have signaled to the EPA that lead contamination was occurring, the combined information available to Region 5 painted a picture of a system at risk from multiple angles.

By compiling and examining these factors, Region 5's staff and managers could have intervened sooner after the source switch, as evidence of risk grew. Instead, the federal response was delayed while residents continued to be exposed to lead in their drinking water.

Risks at the State Level

In addition to the disinvestments previously mentioned, in 2016, EPA managers discovered that MDEQ personnel did not establish water quality parameters required by the LCR.[17] As stated in Chapter 1, the parameters define a range of water chemistry constraints to protect infrastructure from corrosion. The LCR requires water systems to use lead tap and water quality parameter sampling results to verify corrosion control treatment programs are effective. When water samples demonstrate that a system does not meet the parameters, violations can result.

Risks at the Local Level

Despite known economic challenges in the city, EPA Region 5 staff did not identify Flint's source water switch as an event that could impact the city's ability to comply with the SDWA. Beginning in May 2014, a number of problems emerged that should have warned the agency about the risk.

For example, in June 2015, an EPA Region 5 scientist drafted an interim report that outlined concerns about lead in Flint's drinking water and the lack of corrosion control treatment.

This report indicated the potential for serious human health risks and recommended potential EPA actions.

Even with this memo and other information, a senior regional official told the Regional Administrator for Region 5 that the region and state were in the process of reaching an agreement that would require Flint to implement corrosion control treatment.

According to the Regional Administrator, the problem was not presented in a way that conveyed the gravity of the situation. The Regional Administrator said that no wide-spread public health issues were apparent at the time. This delayed the EPA's formal intervention. Table 2 shows the growing evidence of risk in Flint.

---

[17] The LCR requires that water quality parameters include measures for lead, copper, pH, alkalinity, calcium, conductivity, orthophosphate or silica, and temperature. The EPA discovered the lack of water quality parameters in Michigan during the agency-led program review of the state's drinking water program: *U.S. Environmental Protection Agency Public Water System Supervision Draft Program Review of the Michigan Department of Environmental Quality Drinking Water Program 2016*.

Table 2. Evidence of growing risk in Flint

May 2014	• EPA Region 5 learned of Flint drinking water source switch. • EPA Region 5 received first citizen complaints about drinking water quality.
August 2014	• First total coliform violation occurred.
December 2014	• First total trihalomethanes violation occurred.
February 2015	• EPA Region 5 learned of elevated lead level in Flint resident's drinking water. • EPA Region 5 manager voiced concerns to colleagues that sampling protocol is biasing lead results.
March 2015	• Engineering firm contracted to work on the Flint water system issued report that recommends the addition of corrosion control treatment. • EPA Region 5 learned that Genesee County was investigating an increase in *Legionella* cases.
April 2015	• MDEQ staff member told an EPA Region 5 manager that the Flint system does not have any corrosion control treatment.
June 2015	• EPA Region 5 learned that at least four homes had lead in drinking water above the action level. • EPA Region 5 manager shared interim report on high lead levels in Flint's drinking water with an affected resident. The report was subsequently released to the public.

Source: OIG analysis of EPA Region 5 records.

Between May 2014 and the issuance of the EPA Emergency Administrative Order in January 2016, EPA Region 5 staff received 87 citizen complaints about drinking water conditions in Flint. The complaints described distress about water quality (color and odor), more specific concerns about total trihalomethanes, total coliform and *E. coli*, *Legionella* and, ultimately, lead.

Flint Water Treatment Plant (EPA OIG photo, taken April 2016).

Of the complaints received before the order, 30 (34.5 percent) included concerns about lead. Even so, Region 5 staff did not identify the volume of complaints as indications of unusual problems in Flint's water system. Region 5 officials responded with form letters that recommended citizens resolve their concerns by contacting the MDEQ or Flint's water system staff. In six cases, Region 5's response came more than a year after the citizen made the complaint. In 11 cases, we found no documentation of any response.

In April 2016, we interviewed a sample of the complainants. Some expressed frustration at the lack of resolution of their complaints by Region 5. Staff and managers in Region 5 did not have a system for cataloguing and responding to citizen complaints, nor did they use citizen complaints or the volume of calls as indicators of problems with Flint's water system. As a result, Region 5 could not assess the severity of the situation, and staff were not able to alert management to an emerging incident.

Multilingual sign directing residents to a water distribution site in Flint. (EPA photo)

Region 5 staff and managers told us that these risk factors individually or taken together would not have indicated elevated lead levels. They stated, "Generally, there is little or no correlation between citizen complaints about water and lead content." Region 5 staff said residents often complain about aesthetic concerns that are unrelated to lead levels. Further, staff told us that aesthetic complaints are common, especially in under-resourced water systems like Flint's.

Region 5's comments regarding citizen complaints contradict the EPA's *Drinking Water Action Plan*, which was issued in November 2016 and stresses the importance of assessing risks at the local level. Issued after the Flint crisis, the plan states that EPA staff should do the following:

Develop priority indicators to identify troubled water systems. In a world of limited oversight resources, it is critical the EPA and primacy agencies become more effective at proactively directing oversight and technical assistance resources to systems that are most likely to face problems or risks that may adversely affect public health. Indicators may also be used to identify systems that are likely to be in violation, and follow-up with the appropriate compliance assistance and enforcement response.[18]

Although EPA Region 5 receives drinking water complaints regularly, we believe that complaints can serve as a critical indicator of potential problems. Instead of referring complainants back to state and local officials to resolve, the region could have taken a more proactive stance. Further, the volume of complaints should have alerted Region 5 to a developing drinking water risk in Flint. A more proactive approach could have enabled the region to respond more swiftly to the contamination incident.

Region 5 Did Not Use Oversight and Monitoring Authorities and Tools to Influence MDEQ

Region 5 used several tools to oversee the MDEQ's handling of contamination in Flint. However, when those approaches did not result in swift action to protect public health, Region 5 did not employ other SDWA authorities at its disposal to require compliance in Flint. Most authorities were not used because regional managers did not recognize them or deem them appropriate. Table 3 provides examples of drinking water oversight authorities available to Region 5, and our analysis of the region's use of those authorities. Table 4 provides examples of oversight tools available to Region 5, and our analysis of the region's use of those tools.

Region 5 conducted 2010 and 2016 reviews of the MDEQ's SDWA program.

EPA staff and management at the Flint Response Center at Kettering University in Flint. (EPA OIG photo, taken April 2016).

[18] EPA Drinking Water Action Plan, November 2016, page 8.

Region 5 senior officials told us that the LCR limited their ability to intervene. For example, they stated that the rule "does not require the primacy agency to notify and/or seek approval from EPA for significant changes to public water systems." Although the LCR does not require the EPA to be notified of the changes, regional staff knew of the source switch soon after it happened. When the region was informed—a year after the switch— that Flint did not have corrosion control treatment, the region could have intervened earlier to verify the changes occurred in accordance with SDWA regulations.

Table 3. Region 5's available oversight authorities in Flint

Authorities		Region 5's use of authorities
Authority to negotiate resolution of grant issues or impose additional conditions (such as withholding funds) if a grantee is in noncompliance or issues cannot be resolved. ***(Authority: SDWA § 1443; Regulation: 40 CFR § 35.115 and 2 CFR § 200.338)***	✗	Region 5 did not place additional conditions on the Michigan public water systems supervision grant, even when issues were identified.
Authority to issue an administrative order or commence a civil action when the state does not act. ***(Authority: SDWA § 1414)***	✗	Region 5 did not use this authority.
Authority to issue a federal corrosion control treatment requirement when the state does not adequately do so. ***(Regulation: 40 CFR § 141.82)***	✗	Region 5 did not use this authority.
Emergency authority to act when a contaminant may present an "imminent and substantial endangerment" to human health, and the appropriate state and local authorities have not acted to protect public health. ***(Authority: SDWA § 1431)***	✗	Region 5 did not use this authority; however, EPA headquarters issued an Emergency Administrative Order on January 20, 2016.
Authority to require information to determine whether a system is acting in compliance with drinking water regulations. ***(Authority: SDWA § 1445)***	✓	Region 5 requested and received information from Michigan about corrosion control treatment and other issues.
Authority to revoke primacy if a state does not meet primary enforcement requirements. ***(Authority: SDWA § 1413)***	✗	Region 5 did not use this authority.

Source: OIG analysis.

Sampling results from two 6-month tests indicated the 90th percentile of the results exceeded the 5 ppb limit at which optimized corrosion control treatment is required, and lead concentrations increased over the course of testing. According to Region 5 officials, these results did not alert the region to widespread problems in Flint's water system. However, if the results of these tests were combined with other issues identified in Flint, this could have signaled problems in Flint's water system and the region could have intervened more forcefully.

Table 4. Region 5's available oversight tools in Flint

Drinking water oversight tools	Region 5's use of oversight tools
Coordinate enforcement efforts. ***(Guidance: EPA Office of Water National Program Manager Guidance)***	✓ Region 5 conducted regular meetings. It was during these meetings that the state expressed a different interpretation from Region 5 regarding the LCR.
Provide technical and scientific support. ***(Guidance: EPA Office of Water National Program Manager Guidance)***	✓ Region 5 provided technical assistance that the state initially was reluctant to accept. According to Region 5, the MDEQ failed to take appropriate actions when the region consistently identified problems with Flint's water switch.
Receive drinking water compliance data. ***(Guidance: EPA Office of Water National Program Manager Guidance)***	✓ Region 5 received correspondence and reports from the MDEQ, which uncovered compliance issues.
Use Drinking Water State Revolving Fund program monies in support of the PWSS program to help achieve the public health objectives of the SDWA. ***(Authority: SDWA § 1452; Regulation: 40 CFR Part 35, Subpart L; Guidance: Drinking Water State Revolving Fund Program Guidance)***	✓ Region 5 awarded Flint a grant for infrastructure improvements after the 2016 crisis.
Conduct compliance reviews of state programs. ***(Guidance: EPA Office of Water National Program Manager Guidance)***	✓ Region 5 conducted 2010 and 2016 reviews of the MDEQ's SDWA program.

Source: OIG analysis.

For example, Region 5 could have done the following:

- Taken enforcement action under SDWA § 1414, which would have required Flint to install corrosion control treatment after notifying the state.
- Issued an order under SDWA § 1431 when the region had evidence there was "imminent and substantial endangerment" to human health, and when the region knew the state did not act.
- Alerted Flint residents about the potential harm to public health (although not required under the SDWA).

The Flint water crisis demonstrates that public health is not protected when EPA regional staff—with multiple warning signs—do not use the agency's SDWA authorities in conjunction with EPA oversight tools.

Conclusion

Effective program management tools provide a foundation for federal programs to achieve important goals like those in the SDWA.

Oversight tools found in the SDWA are designed to protect public health from drinking water contamination, but they are only effective when used.

During the Flint water crisis, Region 5 leadership did not employ many of its SDWA authorities, as the state continued to debate LCR requirements and residents continued to drink potentially contaminated water. Flint drinking water contamination continued, in part, because the public health protection authorities of the SDWA were not used effectively. The EPA and its regional offices must understand their oversight tools and authorities, and not be reluctant to use them to protect public health.

To understand roles and responsibilities, Region 5 should clarify its region-state relationship with the state of Michigan. Region 5 also should establish a schedule for periodically training drinking water program staff and managers to confirm understanding of their roles and responsibilities, and the EPA's authorities for intervention. To improve drinking water oversight, Region 5 should establish a risk assessment protocol to heighten awareness, and establish control activities to properly monitor state programs. Cognizance of roles and responsibilities, as well as EPA authorities, is important not only for Region 5, but also for the EPA at the national level.

Bottled water distribution in Flint was established during response efforts (EPA photo).

Recommendations

We recommend that the Regional Administrator, Region 5:

3. Publicly document clear expectations, roles and responsibilities between the EPA and the state of Michigan in an official document, such as a memorandum of understanding or a supplemental primacy document.
4. Implement a system for regional drinking water staff, managers and senior leaders, which incentivizes staff elevating and managers addressing important and emerging issues in accordance with the EPA's "Policy on Elevation of Critical Environmental and Public Health Issues."
5. Provide the public with all results from EPA reviews of Michigan's Safe Drinking Water Act program, and track the progress of identified corrective actions.

We recommend that the Assistant Administrator for Enforcement and Compliance Assurance, and the Assistant Administrator for Water:

6. Provide regular training for EPA drinking water staff, managers and senior leaders on Safe Drinking Water Act tools and authorities; state and agency roles and responsibilities; and any Safe Drinking Water Act amendments or Lead and Copper Rule revisions.
7. Implement a system to identify management risks in state drinking water programs, including elements such as atypical events, emerging public health concerns, environmental justice concerns and public health analyses.
8. Create a system that tracks citizen complaints and gathers information on emerging issues. The system should assess the risk associated with the complaints, including efficient and effective resolution.
9. Improve oversight by establishing a clear and credible escalation policy for EPA intervention in states. The policy should provide steps the EPA will take when states do not act.

Agency Response and OIG Evaluation

The EPA provided a coordinated response from Region 5, the OW and OECA. They provided planned corrective actions in their response, and provided additional clarifications in a June 5, 2018, meeting with the OIG. Based on the coordinated response and additional clarifications, the EPA provided acceptable planned corrective actions and completion dates for Recommendations 3 through 9.

Region 5 agreed with Recommendations 3, 4 and 5. For Recommendation 3, Region 5 will add clear roles and expectations, and post the annual Michigan PWSS program work plan and end-of-year evaluation on the EPA's website so that the information is available to the public. The projected completion date is October 1, 2018.

To address Recommendation 4, by October 31, 2018, Region 5 plans to incentivize and provide midyear and end-of-year feedback to staff, managers and senior leaders on elevating and addressing issues of concern. To address Recommendation 5, Region 5 noted that the 2010 and 2016 *Michigan Drinking Water Program Review* reports and the FY 2010 through FY 2016 *End-of-Year Evaluation* reports are currently available to the public online. They also plan to finalize a corrective action plan that stems from the 2016 *Program Review Report* and post this plan online by September 30, 2018, and provide updates on progress by April 30 and September 30 each year thereafter.

OECA and the OW agreed with Recommendations 6 through 9. To address Recommendation 6, the EPA has been providing training on the LCR's optimal corrosion control treatment and optimal water quality parameter requirements. The workshops provide a review of LCR requirements and emphasize the tools and authorities drinking water programs can leverage to implement the requirements more effectively. The training has been delivered through in-person workshops at all EPA regions, as well as through special conference sessions.

Since the inaugural workshop in FY 2016, the training has reached approximately 1,300 drinking water professionals (staff and managers) from federal and state drinking water programs, technical assistance providers and water utilities.

Training was completed on May 31, 2018, and training needs will be reassessed by September 30, 2019. OECA and the OW also stated the training will be provided to senior leaders.

To address Recommendation 7, OECA and the OW initiated a work group with participants from both offices and EPA regions. This work group will identify ways to use drinking water data and other information to identify water systems that present, or are likely to present, a significant risk to public health. The work group will develop procedures and strategies by December 31, 2018.

To address Recommendation 8, OECA and the OW committed to either enhancing an existing system, or developing a new system to track and resolve citizen complaints by September 30, 2019.

For Recommendation 9, a work group including OECA, the OW and EPA regions will develop procedures and strategies to provide timely and effective EPA intervention where a state's response to risks is insufficient to protect human health. The EPA plans to make a policy decision by July 31, 2019.

Region 5, OECA and the OW provided supplemental technical comments that the OIG incorporated as appropriate. The EPA's coordinated response is found in Appendix C.

STATUS OF RECOMMENDATIONS AND POTENTIAL MONETARY BENEFITS

Rec. No.	Page No.	RECOMMENDATIONS				
		Subject	Status[1]	Action Official	Planned Completion Date	Potential Monetary Benefits (in $000s)
1	15	Establish controls to annually verify that states are monitoring compliance with all Lead and Copper Rule requirements, including accurately identifying tier 1 sampling sites and maintaining continuous corrosion control treatment.	U	Assistant Administrator for Enforcement and Compliance Assurance, and the Assistant Administrator for Water		
2	15	Revise the Lead and Copper Rule to improve the effectiveness of monitoring and corrosion control treatment protocols.	R	Assistant Administrator for Water	2/28/19	
3	27	Publicly document clear expectations, roles and responsibilities between the EPA and the state of Michigan in an official document, such as a memorandum of understanding or a supplemental primacy document.	R	Regional Administrator, Region 5	10/1/18	
4	27	Implement a system for regional drinking water staff, managers and senior leaders, which incentivizes staff elevating and managers addressing important and emerging issues in accordance with the EPA's "Policy on Elevation of Critical Environmental and Public Health Issues."	R	Regional Administrator, Region 5	10/31/18	
5	28	Provide the public with all results from EPA reviews of Michigan's Safe Drinking Water Act program, and track the progress of identified corrective actions.	R	Regional Administrator, Region 5	9/30/18	
6	28	Provide regular training for EPA drinking water staff, managers and senior leaders on Safe Drinking Water Act tools and authorities; state and agency roles and responsibilities; and any Safe Drinking Water Act amendments or Lead and Copper Rule revisions.	R	Assistant Administrator for Enforcement and Compliance Assurance, and the Assistant Administrator for Water	9/30/19	
7	28	Implement a system to identify management risks in state drinking water programs, including elements such as atypical events, emerging public health concerns, environmental justice concerns and public health analyses.	R	Assistant Administrator for Enforcement and Compliance Assurance, and the Assistant Administrator for Water	12/31/18	

(Continued)

Rec. No.	Page No.	RECOMMENDATIONS				
		Subject	Status[1]	Action Official	Planned Completion Date	Potential Monetary Benefits (in $000s)
8	28	Create a system that tracks citizen complaints and gathers information on emerging issues. The system should assess the risk associated with the complaints, including efficient and effective resolution.	R	Assistant Administrator for Enforcement and Compliance Assurance, and the Assistant Administrator for Water	9/30/19	

C = Corrective action completed.
R = Recommendation resolved with corrective action pending.
U = Recommendation unresolved with resolution efforts in progress.

APPENDIX A

Table A-1. Timeline of Key Events

Date	Event
Pre-2014	
1967	Flint switched from the Flint River as its primary water supply source to Lake Huron water supplied by Detroit.
2014	
April 17	The Flint water system water quality manager tells the MDEQ that if drinking water from the Flint River is distributed within the next couple of weeks, it will be against his direction.
April 25	City of Flint switched from purchasing water from Detroit to using the Flint River as a water source and using the Flint plant for treatment.
April 30	The MDEQ requires Flint to conduct two 6-month rounds of monitoring for lead and copper (July–December 2014 and January–June 2015) to determine whether corrosion control treatment is necessary.
May	Region 5 staff member first learns about Flint's source water switch from an MDEQ employee. He provided analysis that lead levels were in compliance when leaving the plant.
May 15	First record of citizen complaint to Region 5 after the change in water source. The complaint was about the poor quality of Flint's drinking water.
June 17	An MDEQ staff member emails a Flint water treatment plant operator to confirm that no orthophosphate (a corrosion inhibitor) is to be added so no monitoring is necessary.
July 1	Flint begins first 6-month monitoring period for lead and copper.
August	City of Flint violates the National Primary Drinking Water Regulations (NPDWRs) for E. coli bacteria and total coliform. The city issues a boil advisory on August 15.
Aug. 15	Flint increases flushing of water and boosts chlorine disinfectant use.
Oct. 17	Genesee County Health Department expresses concerns about a Legionellosis outbreak and possible connection to water supply.
Dec.16	The MDEQ notifies Flint of initial quarterly violation of SDWA Disinfection Byproducts (total trihalomethanes) requirements. Subsequent notices of violations were issued in March and June 2015.
Dec. 30	Flint completes its first 6-month round of lead monitoring with results above the PQL of 5ppb.
2015	
Jan. 1	Flint begins the second 6-month monitoring period for lead and copper.

Date	Event
Jan.12	Detroit offers to reconnect Flint's water system to its water system and waive the $4 million connection fee. Flint declines the offer on January 29, 2015.
Feb. 10	Genesee County Health Department epidemiologist consults outside experts about 47 Legionellosis cases, and a possible connection with the Flint water system.
Feb. 16	Flint posts its Flint water frequently asked questions, and states that Flint water is safe to drink.
Feb. 18	Lead levels of 104 ppb found in tap water of a Flint resident's home.
Feb. 25	The EPA assigns an EPA disinfection by-products expert to participate on Flint's Technical Advisory Committee.
Feb. 26	Region 5 staff member receives a call from a Flint resident regarding high lead levels in the water in her home. The resident said the city of Flint and the MDEQ attributed it to premise plumbing, but the resident has plastic pipes in her home.
	Region 5 notifies the MDEQ of high lead levels in tap water of Flint resident. Test results show lead levels at 104 ppb. The MDEQ advises Flint to collect a follow-up sample from the resident's home.
	The sentiment at MDEQ staff discussions: Flint is not above the 90th percentile of 15 ppb for lead, so "not sure why Region 5 (EPA) sees this one sample as such a big deal."
Feb. 26–27	Region 5 scientist emails various EPA staff members, an expert from the EPA's Office of Research and Development, and the MDEQ to inquire about the type of corrosion control treatment (CCT) now used in Flint.
Feb. 27	MDEQ manager informs Region 5 that Flint has an optimized corrosion control program. He also shares the results of the first 6-month round of testing to determine whether the system needs CCT. The 90th percentile of the results were 6 ppb, which is over the PQL of 5 ppb. The result is an indication that CCT is necessary.
	Region 5 scientist voices concerns to EPA colleagues that Flint's lead- sampling protocol (pre-flushing) is biasing lead results toward the low side.
Mar. 3	A follow-up sample of tap water collected by the city of Flint had a lead level of 397 ppb at the same Flint residence where the lead level had been 104 ppb when a sample was first taken on February 18.
Mar. 10	Region 5 staff member emails the MDEQ saying that she has been inundated with citizen emails referred to her from the White House about Flint's water quality problems.
Mar. 26	EPA learns that Genesee County is investigating an increase in Legionella cases.
April 7	Region 5 scientist emails other Region 5 and EPA Office of Research and Development staff saying Flint needs help, and the system needs to be evaluated. The city's practice of pre-flushing ahead of collecting samples results in an inadequate capture of lead released in the service lines.
April 24	An MDEQ staff member tells Region 5 scientist that the Flint system does not have any CCT.
April 25	Region 5 scientist notifies MDEQ staff that CCT should have been maintained with the source switch. Region 5 scientist also tells the MDEQ that lead levels may be much higher than the compliance results indicate due to lack of CCT and pre-flushing.
April 27	Region 5 scientist meets with Flint resident to review internal plumbing, and provide sampling bottles to take water samples for lead in the resident's home.
May 1	The MDEQ informs Region 5 that it had not made a formal decision as to whether the city meets the exemption criteria for corrosion control treatment. The MDEQ also reasoned that since Flint would be switching source water in another year to the Karegnondi Water Authority (KWA), requiring the study at this time would be of little value.
May 6	EPA Region 5 visits a Flint resident's home (see Feb. 18, March 3 and April 27 entries) to collect pipe samples from the service line. Three sections of the service line were extracted and sent to Virginia Tech University for analysis. The EPA leaves sample bottles for the resident to collect sequential samples following the expected replacement of the service line.
	EPA Region 5 collects a set of sequential samples from two additional Flint residences, one had low lead levels, the other high lead levels.
	The city of Flint tests the water at two more homes. Both homes had high lead results.
May 11	Region 5 staff member asks an EPA Office of Research and Development expert about the value of starting CCT immediately. The expert notes that Flint has multiple problems, and CCT should be integrated into a long-term plan.
May 13	Region 5 scientist samples the Flint residence (see Feb. 26 and Mar. 8 entries) after disconnection from an old service line. Results show that water is below the action level.
June 10	During a conference call, Region 5 tells the MDEQ that Flint is likely to need corrosion control treatment, and the expertise of the Office of Research and Development should be used to avoid problems moving forward.

Table A-1. (Continued)

Date	Event
June 24	Region 5 had information that at least four homes in Flint had lead in drinking water in concentrations above the action level.
	Region 5 scientist releases interim report on high lead levels in Flint drinking water.
July 10	Region 5 Regional Administrator writes to Flint's Mayor and states the EPA will work with the MDEQ on issues related to lead in the water.
July 21	Region 5 informs the MDEQ about the region's interpretation of the LCR, and that Flint should have been maintaining corrosion control treatment since the source switch. The MDEQ disagreed with the interpretation, stating that it was premature, and requests a legal opinion. However, the MDEQ did agree to start CCT as soon as possible.
	The MDEQ will not change its sampling protocol regarding pre-flushing until new state regulations are issued.
July 24	The MDEQ emails the EPA the results from the second 6-month monitoring period. The 90th percentile result is 11 ppb, above the PQL of 5 ppb.
Aug. 17	The MDEQ orders Flint to optimize CCT within 6 months.
Aug. 23	Independent researcher notifies city of Flint that his research team has started collecting samples for a water quality study in Flint.
Sept. 3	Flint Mayor announces that CCT will be implemented and invites EPA corrosion control experts to join Flint technical advisory committee.
Sep. 8	Independent research team issues a report indicating that 40 percent of Flint homes have elevated lead levels. "Flint has a very serious lead in water problem."
Sep. 10	OECA Water Enforcement Division (EPA headquarters) attorney acknowledges learning about the Flint situation and articulates plans to follow up with Region 5 staff.
Sep. 11	Local doctor convenes study of children's blood lead levels following findings about lead in drinking water from independent research team.
Sep. 15	Region 5 Regional Administrator calls the MDEQ and Flint Mayor to urge action on drinking water.
Sep. 24	Study completed by local doctor is released showing the numbers of children with elevated blood lead levels increased 90 percent after the switch to Flint River.
Sep. 25	Flint issues formal health advisory regarding lead in the drinking water.
Sep. 27	Region 5 Regional Administrator calls the MDEQ to expedite corrosion control treatment and to provide bottled water to residents.
Sep. 28	Region 5 Regional Administrator emails a 10-Point Plan to the EPA Administrator and asks the Office of Research and Development Director to encourage the MDEQ to call in Michigan Department of Health and Human Services for assistance.
Oct. 1	Michigan's Chief Medical Officer confirms the results from the local doctor's blood lead level study.
	Genesee County Health Department issues do not drink advisory.
	Private and public sources donate $105,000 to fund 5,000 water filters for Flint residents to be distributed to those with the highest risk first.
Oct. 5	EPA learned that the city of Flint claimed to use lime softening as CCT.
Oct. 6	The Technical Advisory Committee recommends a return to Detroit water. The MDEQ and the EPA agree to additional phosphate treatment for the already finished Detroit water.
Oct. 8	Michigan announces that Flint will go back to Detroit water system for drinking water.
Oct. 9	Michigan Congressman requests assistance from EPA Administrator.
Oct. 16	Flint returns to using treated water from Detroit.
	The EPA established the Flint Safe Drinking Water Task Force to provide the agency's technical expertise through regular dialogue with designated officials from the MDEQ and the city of Flint.
Oct. 18	The MDEQ's Office of Drinking Water and Municipal Assistance Director issues statement regarding how his office was mistaken in how it interpreted federal rules governing CCT.

Date	Event
Oct. 23	The EPA's Flint Safe Drinking Water Task Force provides the MDEQ with technical comments on Flint's CCT plan.
Nov. 3	The EPA's Office of Water issues a memo that clarifies the agency's interpretation that the LCR requires large water systems such as Flint to have optimized technologies in place.
Nov. 25	The EPA's Flint Safe Drinking Water Task Force releases its preliminary assessment, with recommendations on sampling protocol, operating procedures, corrosion control, and outreach during transition to the KWA.
Dec. 9	Flint begins supplemental corrosion control treatment.
Dec. 14	Mayor of Flint declares state of emergency.
2016	
Jan. 5	Michigan Governor declares state of emergency for Genesee County due to health and safety issues caused by lead in Flint's drinking water.
Jan. 12	Michigan Governor requests assistance from Federal Emergency Management Agency and activates the Michigan National Guard to help distribute bottled water.
Jan. 14	Michigan Governor requests major disaster and emergency declaration and federal aid.
Jan. 15	Michigan Attorney General initiates investigation to determine whether state laws were violated.
Jan. 16	President declares federal state of emergency for city of Flint with $5 million in aid.
Jan. 21	The EPA issues Emergency Administrative Order to the state of Michigan, the MDEQ and the city of Flint.
Nov. 17	Region 5 acting Regional Administrator issues First Amendment to Emergency Administrative Order.

Source: OIG-created timeline of events. The timeline ends on the date the First Amendment to Emergency Administrative Order was issued. For all subsequent EPA activity, see the EPA's Flint Drinking Water Documents webpage.

APPENDIX B

Status of Emergency Administrative Order and Amendment

Emergency Administrative Order, January 21, 2016

The SDWA provides the EPA with the authority to order actions when there is imminent and substantial endangerment to human health, and when actions taken by state or local authorities are inadequate to protect public health. The EPA determined that responses from the city of Flint and the MDEQ were inadequate, and the EPA issued the SDWA emergency administrative order to the state of Michigan, the MDEQ and to Flint. Paragraphs 1–49 contain the EPA's findings and conclusions. Paragraphs 50–64 identify actions required by the order, with assigned completion dates where applicable.[19] The EPA provided the status of actions on May 30, 2018.[20]

[19] Text is not verbatim.

[20] Region 5's Office of Compliance and Enforcement tracks progress on the emergency order and amendment.

Paragraph Number	**Due Date***	**Required Action**	**Status**
50	1/22/16	Within one day of the effective date of this Order, Respondents[21] shall notify EPA in writing of their intention to comply with the terms of this Order. For the purposes of this Order, "day" shall mean calendar day.	Complete
Reporting Requirements			
51	1/26/16 and weekly thereafter	Within five days of the effective date of this Order, the State shall create, and thereafter maintain, a publicly available website. Respondents must post on this website all reports, sampling results, plans, weekly status reports on the progress of all requirements and all other documentation required under this Order. The Respondents shall not publish to this website any personally identifiable information.	Complete Weekly status reports are ongoing
52	1/31/16	The Respondents shall within 10 days of the effective date of this Order respond in writing, in accordance with Paragraph 51, to all of the EPA Flint Task Force's requests and recommendations made on November 25, 2015 and subsequent dates. The response shall include all actions Respondents have taken and intend to take in response to those requests and recommendations. The EPA Flint Task Force's requests and recommendations are publicly available at http://www.epa.gov/mi/flint-drinking- water-documents	Ongoing – Respondents continue to respond in writing to all of the Technical Support Team's (TST) (f/k/a EPA Flint Task Force) recommendations, including the TST's 1-22-2018 recommendation relating to the implementation of the American Water Works Association 810-17 Standard for replacement and flushing of lead service lines. All of the TST's recommendations and the corresponding City/State responses are on EPA's website.
53	1/31/16	Within 10 days of the effective date of the Order the Respondents shall provide the following information in accordance with Paragraph 51: a. Water quality parameter measurements (pH, total alkalinity, orthophosphate, chloride, turbidity, iron, calcium, temperature, conductivity) in the distribution system. The City is required by the MDEQ permit to monitor for these parameters at 25 sites quarterly and at 10 of these sites weekly;	Complete
	1/31/16	b. All lead in water testing results for the City since January 2013, including those not used for LCR compliance; and	Complete
	1/31/16	c. Identification of areas (e.g., zip codes, neighborhoods) in the City with elevated blood lead levels.	Complete

[21] Respondents are the state of Michigan, the Michigan Department of Environmental Quality, and the city of Flint.

Paragraph Number	Due Date*	Required Action	Status
54	1/31/16	Within 10 days of the effective date of the Order, the Respondents shall provide, without publicly disclosing any personally identifiable information, the following directly to the EPA in accordance with Paragraph 66: a. Existing inventory of homes with lead service lines in Excel or similar format	Ongoing – An initial existing inventory was submitted; however, the inventory was severely lacking in completeness and accuracy. Flint recently submitted an updated estimate of their inventory based on Flint Action and Sustainability Team (FAST Start) project findings
		b. Addresses of homes that have had water service interruptions or street disturbances (e.g., water main breaks, road/sidewalk construction, etc.) within the last year; and	Complete
		c. Addresses of currently unoccupied homes.	Complete
55	Ongoing	Respondents shall cooperate with EPA as the Agency conducts LCR sampling and other diagnostic activities in the City.	Ongoing
Treatment and Source Water			
56	~	To Ensure that treated water meets finished water quality goals and is consistently maintained throughout the distribution system, that existing and potential plant operational and mechanical start-up issues are identified and addressed, and that water plant operations staff are proficient in treating the existing and new source water, Respondents shall comply with Paragraphs 57, 58 and 59.	Ongoing- Flint distribution system water quality monitoring reports are posted on the State of Michigan's Flint website. Current certified plant operator duties are being performed by a contractor. The operator certification level of this contractor meets the state requirements for operating the Flint water treatment plant. Two current employees of the Flint water plant took the D-1 Operator Certification Exam in early May 2018—still awaiting results.
57	Ongoing	Respondents shall maintain chlorine residual in the distribution system in accordance with SDWA and the National Primary Drinking Water Regulations ("NPDWRs")	Ongoing- The City's contractor worked with the City and MDEQ to increase residual monitoring to 25 sites from 10 sites. City continues to sample at these locations to verify that an adequate chlorine residual is present throughout the distribution system. MDEQ approved the City's updated Revised Total Coliform Rule sampling plan on 5-14-2018 to include the required 25 sampling sites.
58	Ongoing	The City shall continue to add corrosion inhibitors (e.g., orthophosphate booster) at levels sufficient to re-optimize corrosion control in the distribution system.	Ongoing- The City continues to add orthophosphate as a corrosion inhibitor with 3.1 mg/L orthophosphate WQP requirement. There was an LCR violation and public notice triggered in January 2017. Orthophosphate data kept current on the State's website. Ongoing corrosion control studies will further inform the City on corrosion control treatment.

(Continued)

Paragraph Number	Due Date*	Required Action	Status
59	2/4/16	To address optimization of corrosion control for the system as operated with its current water source, within 14 days of the effective date of this Order the Respondents shall submit to MDEQ and post in accordance with Paragraph 51:	
		a. Submit a plan and schedule to the MDEQ to review and revise as needed designated optimal corrosion control and water quality parameters as well as monitoring plans for LCR compliance and all other monitoring plans developed to ensure that the treatment plant is consistently and reliably meeting plant performance criteria and all other NPDWRs;	Ongoing - WQPs were submitted, but EPA continues to work with the Respondents to ensure that all the requirements of paragraph 59 are fully met. EPA is currently working with the Respondents to ensure that an LCR Compliance Plan is in place before the City takes over LCR compliance monitoring from MDEQ.
		b. Submit a sampling plan for daily monitoring of water quality parameters in the distribution system with results compiled in a weekly report in an approved format; and	Ongoing – WQP monitoring is being conducted by the City, but EPA continues to work with the Respondents on the development of a formal monitoring plan. The City is in the process of developing this plan.
		c. Submit an operations plan for the corrosion control equipment (storage day tanks, feed/injection systems), with results compiled in a weekly format, that includes monitoring, calibration, verification (pump catch, etc.) as well as daily monitoring of finished water corrosion control parameters. Results shall be submitted and posted weekly.	Ongoing – The corrosion control equipment plan initially submitted was inadequate. EPA, MDEQ, and the City continue to discuss how to finalize this plan. The City's contractor continues to work on related Standard Operating Procedures (SOP).
60	Ongoing	Respondents shall not effectuate a transition to a new water source for the City's PWS (e.g., from KWA) until such time as they have submitted a written plan, developed through consultation with appropriate experts and after providing adequate advanced notice and an opportunity for public comment, to MDEQ and in accordance with Paragraph 51, demonstrating that the City has the technical, managerial and financial capacity to operate its PWS in compliance with SDWA and the NPDWRs and that necessary infrastructure upgrades, analysis, and testing have been completed to ensure a safe transition. Such plans shall include, but not be limited to, provisions addressing:	Revised in the First Amendment to the Emergency Order issued on November 17, 2016 The City has agreed to a 30- year long-term water source, blend of Great Lakes Water Authority (GLWA)/ Genesee County Drainage Commissioner (GCDC) water agreement. See section 60 in the Amendment below for details.
		a. The impacts on corrosion control for any new source water and an operations plan for periodic use of existing sources of water;	
		b. Completion of corrosion control study for any new sources;	

Paragraph Number	Due Date*	Required Action	Status
Treatment and Distribution System Management			
60		c. Implementation of a "performance period" that allows for the demonstration of the adequacy of treatment of the new water source to meet all NPDWRs before it can be distributed to residents; and	
		d. The City's technical, managerial and financial capacity to meet SDWA's applicable requirements, including the NPDWRs, during and after the transition to any new water source.	
61.	2/5/16	Within 15 days of the effective date of this Order, the City must demonstrate, and the MDEQ and State must ensure, the City has the necessary, capable and qualified personnel required to perform the duties and obligations required to ensure the PWS complies with the SDWA and the NPDWRs.	Ongoing –Hiring additional Public Water System staff remains a challenge for the City of Flint because of fiscal limitations. The City now meets operator-in-charge, laboratory supervisor, and technician needs through the use of a contractor. Two current staff from the water treatment plant are scheduled to take the Operator Certification exam in May 2018. EPA continues to request hiring updates during its monthly meetings with the Respondents to ensure that the plant is properly staffed.
62.	2/21/16	To ensure the City's PWS is adequately operated to meet SDWA and all NPDWRs, within 30 days of the effective date of this Order, the Respondents shall submit the steps they will take to develop and implement a distribution system water quality optimization plan to MDEQ and in accordance with Paragraph 51, to evaluate and improve its programs that affect distribution system water quality, including: evaluating conditions within the distribution system; creating better documentation; and enhancing communication between the various utility functions that impact distribution system water quality. The MDEQ must ensure that this plan is adequate to ensure SDWA compliance and the State must ensure it is executed.	Ongoing— An inadequate plan was submitted in 2016. The City's contractor recently updated the distribution system optimization plan, and this document is expected to be finalized soon. Ongoing work on SOP development continues.

(Continued)

Paragraph Number	Due Date*	Required Action	Status
Independent Advisory Panel ("IAP")			
63.	1/28/16	Within seven days of the effective date of this Order, the MDEQ and State, with the City's input and concurrence, shall engage a panel of independent, nationally recognized experts on drinking water treatment, sampling, distribution system operation, and members of the affected community to advise and make public recommendations to the City on steps needed to mitigate the imminent and substantial endangerment to the health of persons and general operation of the City's PWS to ensure compliance with SDWA and the NPDWRs.	Panel was engaged and is known as the Flint Water Interagency Coordination Committee (FWICC). However, no public recommendations have been made. NOTE: Frequency of meetings reduced from monthly to every other month. Only one meeting has occurred in 2018.
64.	Ongoing	The charge to the IAP will include the following: a. Make recommendations to the Respondents, and for consideration by the EPA, to ensure the safe operation of the City's PWS.	Ongoing See Michigan governor's website. One resolution was passed, but was not a recommendation. It was to extend the declaration of the Flint Water Emergency. However, MDEQ frequently refers to "FWICC Recommendations" and then says they are not "official." No recommendations have been made for consideration by EPA.
		b. Make other recommendations to the Respondents, and for consideration by the EPA, to better serve the community served by the City's PWS.	See above

* Stated as within number of days since the effective date of this order—January 21, 2016. Note: For current status and details, see the MDEQ's Taking Action on Flint webpage.

First Amendment to Emergency Administrative Order, November 17, 2016

Issued by Region 5, the First Amendment to Emergency Order clarifies what is expected from the state of Michigan, the MDEQ and the city of Flint. The amendment is intended to provide flexibility for respondents to develop their own plan while creating accountability to provide safe and reliable drinking water. The amendment also was designed to restore public confidence.

Required Actions	Due Date	Status
Paragraph 60 in the January 21, 2016, Order Shall Now Read:		
a. Confirmation of Water Sources		
The City shall confirm in writing to EPA its intended new water source and emergency back-up water source within five days of the effective date of the First Amendment. Nothing in this First Amendment prevents the City from identifying a different new water source. The City must notify EPA in writing within five days if there are any changes in its initial, or any subsequent, new water source designation.	11/22/2016 Extension granted to 12/1/2016	Completed- On 11/30/2016. Initial notification of recommended source change received from Mayor on 4/19/2017 – staying on GLWA and using GCDC-treated KWA water as emergency back- up. The City Council entered into a long-term water supply contract with GLWA on 11/21/2017.
b. Development and Implementation for New Water Source Treatment		
i. Pipeline Plan It is necessary to complete the KWA pipeline connection to the Flint water treatment plant ("WTP"). Respondents shall submit to MDEQ for its review and approval, to EPA for its review, and post to the public website under Paragraph 51, a written plan for completing the KWA pipeline connection to the Flint WTP ("Pipeline Plan"), within twenty-one days of the effective date of this First Amendment. The Pipeline Plan shall specify dates for major milestones, including at a minimum, the following:.	12/8/2016	This requirement is no longer applicable due to the City choosing to continue to rely on GLWA as its primary water source. However, the City still needs to complete the pipeline to connect to its backup source, GCDC.
Complete engineering drawings; Submit permit applications and receive approvals; Request contract bidding and award; and Develop construction time table, including final completion date		
Upon MDEQ's written approval of the Pipeline Plan, Respondents shall implement the Pipeline Plan, which must provide for pipeline completion and operation at least three months before the planned distribution date from any new water source.		
ii. Water Treatment Plant Modification Plan ("WTPMP") Respondents shall submit to MDEQ for its review and approval, to EPA for its review, and post to the public website under Paragraph 51, by February 1, 2017, a written WTPMP that provides a preliminary evaluation for Flint's treatment of its identified new source water. The WTPMP shall include: A. An assessment of the treatment processes for the new source water; B. Identification of necessary Flint WTP infrastructure improvements, including the assessment of associated operation and maintenance needs; and C. A schedule with completion dates for major milestones, including, at a minimum, the following: (1) identifying, securing and utilizing funding source(s) and (2) implementing the necessary infrastructure upgrades and other identified improvements. Upon MDEQ's written approval of the WTPMP, Respondents shall implement the WTPMP.	2/1/2017	Submitted- On 3/1/2018 the City of Flint submitted the Water Treatment Plant Modification Plan to EPA and MDEQ. The document is currently under review.

(Continued)

Required Actions	Due Date	Status
iii. New Source Treatment Plan ("NSTP") Respondents shall submit to MDEQ for its review and approval, to EPA for its review, and post to the public website under Paragraph 51, as soon as available and no later than the dates set forth below, a written plan to treat the new source water. The NSTP shall address the City's technical, managerial, and financial capacity to operate its PWS in compliance with the SDWA and NPDWRs, including requirements for optimal corrosion control treatment and water quality parameter monitoring. The NSTP shall be developed in consultation with appropriate experts and the public through adequate advanced notice and opportunity for comment. Prior to submittal of the NSTP, Respondents shall develop a corrosion control study for the new source water and submit the study to MDEQ for its review and approval, and to EPA for its review, by February I, 2017. The NSTP shall be submitted by March 1, 2017, and shall specify a schedule with completion dates for major milestones, including, at a minimum, the following: A. Finalizing necessary standard operating procedures ("SOPs") for each aspect of the water treatment process for the Flint WTP;	2/1/2017	In Progress- The City completed a "coupon study" (not requested or required by EPA) on 11-20-2017 to further inform the pipe loop corrosion control study (required by EPA). The City is currently conducting Phase 1 of the pipe loop study for the current GLWA primary source, despite the TST urging the City to bypass Phase 1 and go directly to Phase 2. The City will also conduct
Implementing infrastructure upgrades that were identified under the WTPMP; Conducting a corrosion control study for the new source water, including the analysis and testing of the impacts on corrosion control treatment under various circumstances to ensure a safe transition; and Developing and implementing a "performance period," which shall begin after the completion of the KWA pipeline connection to the Flint WTP, addressed in Paragraph 60(b)(i), and after the completion and implementation of all applicable requirements in Paragraph 60(b)(ii) and (iii). The performance period shall last as long as necessary, but no less than three months, to allow for the demonstration of the adequacy of treatment of the new water source to meet all SDWA and NPDWRs before it can be distributed to consumers.		Phase 2 of the pipe loop study, but has suggested it focus only on the premise plumbing and not lead service lines, and will not consist of the final source water blend of GLWA and GCDC. EPA continues to discuss how to move forward and complete this study with Respondents. Unable to be completed until Flint has an approved New Source Treatment Plan that it is implementing and until Flint has been approved to start distribution of new source water. Flint is currently working on its New
Upon MDEQ's written approval of the NSTP, Respondents shall implement the NSTP. After completion of the approved NSTP, and at least five days before the proposed distribution of the new source water, Respondents shall: (1) certify in accordance with Paragraph 60(c)(iv) that all elements of the NSTP have been implemented and (2) notify the public in accordance with Paragraph 51.		Source Treatment Plan and will not be ready to distribute new source water until late 2019 at the earliest.

	Required Actions	Due Date	Status
	iv. Use of the Current Water Source Respondents must continue to use the current GLWA source to provide drinking water to the City until the City has demonstrated that all requirements of Paragraph 60 are met and EPA has concurred. Respondents shall provide documentation to EPA, and make publicly available under Paragraph 51, within thirty days of the effective date of the First	12/17/2016	Completed- An extension of several months was requested by Mayor Weaver on 12/16/2016. On 11/30/2017, EPA
	Amendment, that Respondents have made arrangements to have continued access to its current GLWA source water until its transition to a new source water is complete.		received the City's signed Master Agreement for the 30-year GLWA contract with the State, GLWA, KWA, and GCDC as other parties to the agreement.
c.	Reporting and Notification Requirements		
	i. Respondents shall provide monthly updates regarding schedules and milestones, including amount of funds committed, by whom, and when funds will be available for disbursement under Paragraph 60, on the 1st day of each month on the public website under Paragraph 51. Respondents shall continue to report monthly until all necessary requirements of Paragraph 60 are met.	By the first of each month	Ongoing- All reports due thus far have been completed and respondents continue to compile and submit reports for future monthly updates.
	ii. If any event occurs, or has occurred, that may delay Respondents' ability to meet any schedule or milestone in Paragraph 60, Respondents shall notify EPA of that event within five days. If Respondents anticipate any reason they may be delayed in meeting any schedule or milestone in Paragraph 60, Respondents shall notify EPA within five days of the date they become aware of that reason for delay.	Within 5 days of Respondent becoming aware of delay	Ongoing
	Required Actions	Due Date	Status
	Within 10 days of providing such notice to EPA regarding a delay in meeting schedules or milestones, Respondents shall provide contingency plans to address each delay to MDEQ for its review and approval and to EPA for its review.	Within 10 days of notifying EPA of delay	
	iii. If at any point the City decides to change its water source specified under Paragraph 60(a), the City shall notify EPA in writing within five days of such decision. All provisions of Paragraph 60 will apply to any change in water source.	Within 5 days of a decision to change Flint's water source	Ongoing - Currently the City has agreed to a 30- year, long-term GLWA/GCDC agreement.
	iv. Respondents shall provide to EPA a written certification, as specified under Paragraph 67 of the Order, each time a plan, schedule, or milestone required under Paragraph 60 is fully implemented, until EPA has concurred that all requirements under Paragraph 60 have been fully implemented.	Each time a plan, schedule, or milestone under this paragraph is implemented	Ongoing

Appendix C: Agency Comments and OIG Evaluation

May 30, 2018

Memorandum

SUBJECT: Response to the April 13, 2018, Office of Inspector General's Draft Report, "Management Weaknesses Delayed Response to Flint Water Crisis"

FROM: Susan Parker Bodine, Assistant Administrator Office of Enforcement and Compliance Assurance

David P. Ross, Assistant Administrator Office of Water

Cathy Stepp, Regional Administrator Region 5

TO: Kevin L. Christensen, Assistant Inspector General Office of Audit and Evaluation

Thank you for the opportunity to respond to the issues and recommendations presented in the Office of Inspector General (OIG) Draft Report (Project No. OPE-FY16-0031) regarding the response to the Flint water crisis. We also appreciate the additional time you provided us, so the Office of Enforcement and Compliance Assurance (OECA), the Office of Water (OW) and Region 5 could review OIG's April 13, 2018, Draft Report and provide this joint response.

OECA, OW and Region 5 agree with all of OIG's recommendations and have already begun undertaking steps consistent with these recommendations. Below we provide our planned corrective actions and projected completion dates. Further, we have included as an attachment an update regarding the status of the Safe Drinking Water Act (SDWA) emergency order EPA issued in the Flint, Michigan, case (Attachment 1).

A critical part of EPA's core mission is to protect public health and ensure the safety of our nation's drinking water. We appreciate OIG's efforts to bring attention to these important issues. The Agency is working actively to improve SDWA compliance and help safeguard human health.

Agency's Response to Report Recommendations

OECA, OW and Region 5 appreciated the opportunity to ask clarifying questions regarding the recommendations during our May 7, 2018 teleconference. The Report makes nine recommendations. As explained during the teleconference, OECA, OW and Region 5 concur with these recommendations and will work expeditiously to implement them.

Recommendation No. 1

OIG recommends OECA and OW establish controls to annually verify that states are monitoring compliance with all LCR requirements, including accurately identifying tier 1 sampling sites and maintaining continuous corrosion control.

OECA and OW concur with this recommendation. OW has recently developed a standardized Annual Program Review template for the Public Water System Supervision (PWSS) program. This management review is conducted by the Regions for each State and covers required elements, including rule implementation. Within this section, specific LCR implementation metrics can be, and often are, discussed. This Annual Program Review template is being implemented in FY 2018, and OW will continue to adjust for future reviews to potentially include requiring an LCR implementation discussion in all future reports. In addition, EPA also conducts file reviews, which are data audits of the State's program files, approximately every three to five years. During a file review, EPA evaluates state implementation of LCR requirements, including tier 1 sampling sites and corrosion control. For compliance monitoring and enforcement programs, OECA will continue to explore options for creating a national oversight approach for the drinking water programs and present recommendations to its senior managers by June 2019.

Supplemental EPA Response and OIG Disposition of Corrective Actions:

The EPA provided supplemental information in a June 5, 2018, meeting with the OIG. In that meeting, and in subsequent correspondence on June 13, 2018, OECA and the OW revised the response herein to state that they agree on the value of national oversight for compliance monitoring and enforcement of the Drinking Water Program, but added that it is vital that they evaluate and develop it in a collaborative manner with their state partners. The offices plan to work with the states to develop an approach or a pilot for implementing a national compliance monitoring and enforcement oversight program for the Drinking Water Program, as appropriate, by June 2019. The OIG did not accept this response as meeting the intent of the recommendation, which asked that OECA and the OW establish annual controls. We do not agree that developing a pilot or approach constitutes establishing annual controls, per the recommendation language. Recommendation 1 is unresolved.

Recommendation No. 2

OIG recommends OW include in the revised LCR the most protective protocol for monitoring and corrosion control.

OW concurs with this recommendation regarding the importance of proper implementation of the protocol for monitoring and corrosion control, and we continue to work on the long-term revisions to the existing LCR. Most recently, OW engaged stakeholders as part of a federalism consultation.

The Agency is evaluating input we received from our state, local and tribal partners as well as the best available peer-reviewed science to ensure the Rule reflects the best ways to improve public health protection.

Supplemental EPA Response and OIG Disposition of Corrective Actions:

During a June 5, 2018, meeting, the OIG agreed to revise this recommendation. The revised recommendation reads as follows:

We recommend that the Assistant Administrator for Water:

2. Revise the Lead and Copper Rule to improve the effectiveness of monitoring and corrosion control treatment protocols.

This revision takes into account information the EPA must consider in developing the revised Lead and Copper Rule. Based on the revision, the OIG accepted this response from the EPA as meeting the intent of Recommendation 2. The recommendation is resolved pending completion of corrective actions.

Recommendation No. 3

OIG recommends Region 5 publicly document clear expectations, roles and responsibilities between the EPA and the state of Michigan in an official document, such as a memorandum of understanding or a supplemental primacy document.

Region 5 concurs with this recommendation. To implement this recommendation, the Region will document clear expectations, roles and responsibilities between the EPA and Michigan in the annual workplan for the PWSS program grant. To implement this recommendation, Region 5 will post the annual Michigan PWSS program workplan and end of year evaluation on EPA's website so that the information is publicly available.

The PWSS grant workplan covers work conducted as part of SDWA primacy/regulatory requirements, voluntary program efforts and EPA/State work sharing. It also includes commitments as well as additional expectations in the form of specific targets and measures, including both national and Regional measures. These extensive performance measures provide additional expectations of the State and assist the Region in evaluating the State's performance.

The new PWSS program workplan for FY 2019 provides an overall summary that identifies state-specific priorities and core program descriptions that serves as a brief overview for senior management. The workplan also provides separate files with the program-specific summaries that have more detail for staff and mid-level management, such as rules/primacy, data management and reporting, enforcement and compliance assistance, laboratory certification and sanitary surveys.

Supplemental EPA Response and OIG Disposition of Corrective Actions:

The EPA provided supplemental information during a June 5, 2018, meeting with the OIG. In that meeting, EPA Region 5 said it will develop a clarified statement of roles and responsibilities. The region will document clear roles and responsibilities in the annual PWSS work plans. Based on the revisions, the OIG accepted this response as meeting the intent of Recommendation 3. The recommendation is resolved pending completion of corrective actions.

Recommendation No. 4

OIG recommends Region 5 implement a system for regional drinking water staff, managers and senior leaders, which incentivizes staff elevating and management addressing important and emerging issues in accordance with EPA's Policy on Elevation of Critical Environmental and Public Health Issues. Region 5 concurs with this recommendation and has worked to foster an environment where management encourages issue elevation and staff have opportunities to elevate issues of concern to management, especially when there appears to be a substantial threat to public health. On August 15, 2017, Administrator Pruitt reaffirmed to all staff the importance of elevating critical environmental and public health issues so that EPA can properly assess them and respond at appropriate policy and governmental levels in a timely and effective manner. Within Region 5, staff are encouraged to participate and share concerns in a variety of meetings including team, section, branch, division, and regional meetings. These regular check-in meetings among managers and staff allow for information to flow from regional leaders down to staff and vice versa. Region 5 agrees to explore ways to incentivize staff to elevate issues of concern during these engagement opportunities including providing feedback during mid-year and end-of-year reviews with staff, specifically on the customer service critical element present in all staff PARS agreements. In addition to discussion during performance reviews, supervisors have the following award mechanisms available to recognize staff efforts: Agency's annual awards process recognizing key accomplishments, time-off awards, and monetary awards. Region 5 will further explore ways to incentivize staff and management to elevate and address important and emerging issues, such as a peer-to-peer recognition system.

Supplemental EPA Response and OIG Disposition of Corrective Actions:

The EPA provided supplemental information during a June 5, 2018, meeting with the OIG. In that meeting, Region 5 said it will incentivize staff, managers and senior leaders to elevate issues of concern. Region 5 also clarified that it will provide feedback during midyear and end-of-year reviews with staff, managers and senior leaders, and that it will identify ways to further incentivize staff, managers and senior leaders to elevate and address issues. Based on the revisions, the OIG accepted this response as meeting the intent of Recommendation 4. The recommendation is resolved pending completion of corrective actions.

Recommendation No. 5

OIG recommends Region 5 provide the public with all results from EPA reviews of Michigan's Safe Drinking Water Act program, and track the progress of identified corrective actions.

Region 5 concurs with this recommendation and has already made information publicly available. The 2010 and 2016 Michigan Drinking Water Program Review Reports are currently available on EPA's website (https://www.epa.gov/mi). In addition, Region 5's FY 2010 – 2016 End-of-Year Evaluation Reports for the State of Michigan's PWSS Program are available to the public through FOIA on-line (https://foiaonline.regulations.gov/foia/action/public/view/request?objectId=090004d281731e1c)

Region 5 ensures that issues found in previous data and enforcement program reviews of the State's PWSS implementation are included in the following year's annual PWSS program workplan and tracked for progress. Region 5 and MDEQ are finalizing a Corrective Action Plan that has been developed from the recommendations in the 2016 Program Review Report, which Region 5 released in October 2017 (https://www.epa.gov/mi/2016-michigan-drinking-water- program-review). Once final and approved by Region 5, Region 5 will post the Corrective Action Plan to the Agency's website and provide progress updates on a quarterly basis. In addition, the Michigan Corrective Action Plan will be attached to the FY 2018 and subsequent PWSS grant workplans so that both State and EPA commitments are clear. Frequent check-ins with the State are planned to ensure progress in implementing EPA's recommendations to strengthen Michigan's drinking water program.

Beginning with the FY 2019 PWSS grant, the Region 5 Water Division Director will document that EPA recognizes those primacy activities that Region 5 States are not fully implementing and will require the States to provide both short-term and long-term plans for implementing these activities.

OIG Disposition of Corrective Actions:

The OIG accepted this response from the EPA as meeting the intent of Recommendation 5. The recommendation is resolved pending completion of corrective actions.

Recommendation No. 6

OIG recommends OECA provide regular training for EPA drinking water staff, managers and senior leaders on SDWA tools and authorities; state and agency roles and responsibilities; and any SDWA amendments or LCR revisions.

As noted during our May 7, 2018 teleconference with OIG, OECA and OW share responsibility for such trainings and plan to work together to implement this recommendation. Accordingly, they asked that this recommendation be addressed to both OECA and OW. OECA and OW have provided national SDWA training covering a range of topics, including statutory authorities, such as SDWA Section 1431 and Section 1414, and LCR requirements. OECA and OW commit to expanding their training for states, regions and Headquarters on these SDWA issues.

As part of ongoing Agency efforts to enhance national implementation of the LCR, EPA has been providing training on the Rule's optimal corrosion control treatment and optimal water quality parameter requirements. The workshops provide a review of LCR requirements and emphasize the tools and authorities drinking water programs can leverage to implement the requirements more effectively. The training has been delivered through in-person workshops at each of the EPA Regions, as well as through special conference sessions. Since the inaugural workshop in FY 2016, the training has reached approximately 1,300 drinking water professionals (staff and managerial level) from federal and state drinking water programs, technical assistance providers and water utilities.

Supplemental EPA Response and OIG Disposition of Corrective Actions:

The EPA provided supplemental information during a June 5, 2018, meeting with the OIG. In that meeting, OECA and the OW clarified that the training described will also be provided to senior leaders. Based on the revisions, and subsequent correspondence on June 12, 2018, the OIG accepted this response as meeting the intent of Recommendation 6. The recommendation is resolved pending completion of corrective actions.

Recommendation No. 7

OIG recommends OW implement a system to identify management risks in state drinking water programs, including elements such as atypical events, emerging public health concerns, environmental justice concerns and public health analysis.

OW concurs with this recommendation. EPA has initiated a workgroup with participation from OECA, OW and the Regions. The workgroup will explore how best to use drinking water data and measures to identify public water systems that present or are likely to present a significant risk to public health. The workgroup will develop procedures and strategies to ensure timely and effective intervention where risks to public health are identified. Initial findings and the workgroup's proposed framework for action are expected by the end of December 2018.

OIG Disposition of Corrective Actions:

The OIG accepted this response from the EPA as meeting the intent of Recommendation 7. The recommendation is resolved pending completion of corrective actions.

Recommendation No. 8

OIG recommends OW create a system that tracks citizen complaints and gathers information on emerging issues. The system should assess the risk associated with the complaints, including efficient and effective resolution.

OIG Disposition of Corrective Actions:
The OIG accepted this response from the EPA as meeting the intent of Recommendation 8. The recommendation is resolved pending completion of corrective actions.

As explained during our May 7, 2018 teleconference with OIG, OECA and OW also plan to work together to implement recommendation no. 8 and, thus, we ask the OIG to address this recommendation to both offices. OECA and OW are exploring ways to adapt or leverage existing tools. For example, OECA currently supports a public, online Report a Violation (RAV) tool (https://echo.epa.gov/report-environmental-violations) that collects general information about violations. That tool is currently not specific to drinking water data, but it has been used to receive descriptions of potential SDWA violations. EPA may explore possible revisions that would capture more precise contaminant information if reporting capabilities within SDWA, SDWIS Prime, or STORET data systems cannot be modified to receive citizen drinking water contamination data. EPA could use the RAV as the foundation of an improved system that captures citizen reports and sorts them into ranked categories to facilitate identification of management risks that require EPA response.

Recommendation No. 9

OIG recommends OECA and OW improve oversight by establishing a clear and credible escalation policy for EPA intervention in states. The policy should provide steps the EPA will take when states do not act.

OECA and OW concur with this recommendation. On August 15, 2017, Administrator Pruitt reaffirmed EPA's Policy on Elevation of Critical Environmental and Public Health Issues. The Administrator directed EPA staff to elevate concerns quickly and directed the Regions to inform headquarters of any issues that are elevated under this policy.

OECA is providing training on the use of SDWA Section 1431 authority. In implementing the recommendation from the OIG's SDWA Section 1431 Management Alert in October 2016 for OECA to update the 1991 SDWA Section 1431 guidance, over the past year, OECA worked with several Regions, OW and OGC to develop updates to the guidance. OECA is also conducting trainings on Section 1431 and the updated guidance.

OECA is currently considering the possibility of a national initiative to promote improved drinking water compliance. EPA has initiated a workgroup with participation from OECA, OW and the Regions. The workgroup will explore how best to use drinking water data and measures to identify public water systems that present or are likely to present a significant risk to public health. The workgroup will develop procedures and strategies to ensure timely and effective EPA intervention where a state's response to the risk is insufficient to protect the public's health.

OECA will seek state input on whether to create a new national initiative to improve drinking water compliance starting in June 2018, and then will seek public comment in November 2018. OECA expects to make a decision after this engagement process by July 2019.

OIG Disposition of Corrective Actions:

The OIG accepted this response from the EPA as meeting the intent of Recommendation 9. The recommendation is resolved pending completion of corrective actions.

Clarifying Information Regarding Report Findings

EPA appreciates the opportunity to provide some clarifications to inform the Report's findings. EPA has included clarifications below, and Attachment 2 includes a more detailed itemization of clarifications and desired corrections.

First, EPA has identified instances where the Report's simplifications or generalizations result in statements that are factually incorrect or otherwise inconsistent with the Lead and Copper Rule's (LCR or Rule) requirements. EPA recognizes and appreciates that OIG's Draft Report aims to simplify and summarize the LCR's requirements and acknowledges that the Report's findings about the LCR provide important information on EPA's implementation of the Rule.

Second, we want to highlight EPA's concerns about the Draft Report's detailed timeline of events; we think additional information needs to be included for it to be complete and accurate. This timeline, which involves several different government agencies and several EPA offices, is important in understanding the findings and recommendations in the Draft Report. We believe that the additional information provided will allow for a more accurate timeline.

The major clarifications that EPA would like to provide on OIG's Draft Report are as follows:

1. The Report's use of generalized terminology such as "maintaining corrosion control," without including the term "treatment," adds ambiguity to the audit's findings by implying Flint was or could have complied with the LCR without installing treatment,

which is factually inaccurate. There are two intentionally distinct terms used in the LCR regarding corrosion control; these are "corrosion control" and "corrosion control treatment." The LCR provides that systems can be deemed to have optimized corrosion control with or without treatment so there is an intentional distinction between "optimized corrosion control" and "optimized corrosion control treatment," which is especially important with respect to Flint. The term "corrosion control treatment" should be used throughout the Report when referring specifically to the Flint water system.

OIG Response:

The OIG reviewed our use of these terms, and we made changes to the report as appropriate.

2. The Draft Report includes several references suggesting a 5 parts per billion (ppb) benchmark is used to determine when corrosion control treatment installation is required. EPA would like to clarify that all large systems must install corrosion control treatment, regardless of their tap sampling results, unless they meet certain criteria identified in 40 C.F.R. Section 141.81(b).

EPA would also like to clarify that the LCR's treatment technique requirements (e.g., corrosion control treatment installation, lead service line replacement, public education) are triggered when 90th percentile concentrations exceed the lead action level of 15 ppb. Average concentrations are not used, as currently suggested in the Report.

OIG Response:

The OIG reviewed our description of the LCR's use of the PQL, and we made changes to the report to clarify the EPA point listed above.

3. To provide a clear description and timeline of events, additional information is required. Two areas requiring additional clarity are:

A. Instances in which the Draft Report concludes that EPA had access to timely information, when, in fact, EPA believes the state did not provide complete, accurate, and timely data to EPA:

- Page 12, 1st paragraph under "Corrosion Control Was Not Maintained":

> "The MDEQ did not require Flint's public water system staff to maintain corrosion control when Flint changed water sources in April 2014. According to EPA Region 5, the MDEQ concluded that this change in source water would classify Flint as a new drinking water system."

The above sentence conveys that EPA knew of MDEQ's decision to classify Flint as a new water system in 2014; EPA was not aware of this decision until 2015.

- Page 19, 2nd paragraph under "Region 5 Did Not Have an Effective System for Risk Assessment":

> "By compiling and examining these factors, Region 5's staff and managers could have intervened earlier to oversee the drinking water source switch and subsequent events. Instead, the federal response was delayed while residents continued to be exposed to lead in their drinking water."

The above paragraph does not accurately characterize the timeline of events. The Total Coliform Rule and Disinfectants and Disinfection Byproducts Rule maximum contaminant level violations occurred after the source switch. It is unclear how violations that occurred after the source switch could have led Region 5 to oversee something that had already happened (the source switch).

- Page 24, 1st paragraph:

> "Although this is true, regional staff knew of the source switch soon after it happened."

This sentence overlooks the fact that the LCR was specifically amended to require advanced notification of long-term source and treatment changes to the State, and to require the State to review and approve of these changes, not EPA. Region 5 did not learn of the lack of corrosion control treatment maintenance until a year later, on April 24, 2015, when the State communicated that Flint did not have corrosion control treatment.

OIG Response:

The OIG included additional text on page 28 noting that Region 5 did not know about the MDEQ misclassification until 2015. The OIG did not make substantial changes to the discussion of risk assessment. The objective for this discussion is to note that information from various sources came to Region 5 over a period of time—both before Flint changed drinking water sources and after the change occurred. As information about risks mounted over time, Region 5 may not have known that lead contamination existed, but it could have identified Flint as a system with a high level of risk. This could have enabled the region to focus additional attention, questions, assistance, inspections and, where appropriate, enforcement resources on Flint and the MDEQ to forestall a potential drinking water incident like the one that occurred in Flint.

Key activities that Region 5 completed between May 2014 and January 2016 to respond to citizen complaints are not clearly described in the Report. These actions are crucial to understanding how EPA responded to the scope of the recognized risks to public health.

- Page 30, February 25, 2015 - Requested insert for timeline: "EPA assigns Dr. Michael Wright, an EPA disinfection by-products expert, to participate on Flint's Technical Advisory Committee."
- Page 31, April 25, 2015 - Requested insert for timeline: "Region 5 staff notifies MDEQ staff that CCT should have been maintained with the source switch. Moreover, even with MDEQ's incorrect interpretation of the LCR, the July-Dec 2014 LCR compliance results of 6 ppb would have triggered the need for CCT since the 90th percentile was above the threshold in 40 C.F.R. Section 141.81(b)(3)."
- Page 31, April 27, 2015 – Requested addition to timeline: "Region 5 scientists meet with [resident]and provide sampling bottles for her to take water samples for lead in her home."
- Page 31, April 29, 2015 – Requested addition to timeline: "Region 5 meets with and informs the MDEQ of the Region's interpretation of the LCR and that Flint should have been doing corrosion control treatment since the source switch."

OIG Response:

The OIG included additional timeline events where the EPA was able to provide documentation.

4. The Draft Report highlights events that, from OIG's perspective, indicated an impending lead crisis in Flint. However, for the following reasons, at the time the events occurred, while they indicated problems, they were not clear indications of a lead crisis:

- Bacteriological contamination in public water systems is common in the summer months.
- Disinfection by-product levels are typically highest in the summer months and exceedances may occur due to higher water temperatures and increased chlorine levels used in the summer months.
- The residence with the high lead levels in early 2015 was an atypical situation (i.e., not indicative of other homes in the City). The home had an unusually long service line and the service line was physically disturbed in two places. Consequently, although this situation posed an obvious risk to the residents at that home, the high lead levels at that home did not indicate a system-wide concern, especially given that the Michigan Department of Environmental Quality (MDEQ) had indicated that the system had corrosion control treatment in place at that time. It was not until April 2015 that EPA learned from MDEQ that Flint did not have corrosion control treatment in place.
- The LCR's 90th percentile compliance standard allows for 10% of homes to exceed the action level of 15 ppb. The City's 90th percentile values were at 6 ppb

(December 2014) and 11 ppb (June 2015). It was not until September 2015 that EPA learned the City was not sampling at the required tier 1 (highest risk) sites for LCR compliance monitoring.

OIG Response:

The OIG's discussion on risk assessment is intended to describe how information that Region 5 accrued over time demonstrated a system at risk. In addition to the information in the bullets, the region also received citizen complaints, knew that the MDEQ had disinvested from some SDWA requirements, and that the city recently changed drinking water sources. As described above, Region 5 may not have known there was a specific risk of lead contamination, but it could have identified Flint as a system with a high level of risk. The region then could have focused additional attention, questions, assistance, inspections and, where appropriate, enforcement resources on Flint and the MDEQ to forestall a potential drinking water incident like the one that occurred in Flint.

5. The Draft Report refers to the LCR's Section 141.81(b)(3) criteria to allow Flint's system to meet the exemption criteria for corrosion control treatment. However, Flint never met the criteria in (b)(3). Flint's public water system was required to have corrosion control treatment; it met that requirement by the 1997 deadline through purchasing water from Detroit treated with orthophosphate. As EPA noted in its 2016 Michigan Drinking Water Program Review Report (https://www.epa.gov/mi/2016-michigan-drinking-water- program-review), MDEQ did not correctly review and approve Flint's source/treatment change in 2014.

OIG Response:

The OIG made modifications to the description of Flint's sampling under the LCR's Section 141.81(b)(3).

6. There is an important clarification needed to the timeline section on page 30 of the Draft Report. In February 2015, there were three statements in the email provided to Region 5 by an MDEQ manager that led EPA to believe corrosion control treatment had been maintained. MDEQ stated: (1) Flint has an optimal corrosion control program in place; (2) Flint is conducting water quality parameter monitoring; and (3) the system is in compliance with all treatment technique requirements. These statements, taken together, indicated that MDEQ believed that Flint was in compliance with the LCR requirement to maintain corrosion control treatment and that the corrosion control treatment was being monitored using the required water quality parameter monitoring.

OIG Response:

The OIG's description on page 20 provides a summary of this point.

Attachments

OIG Report: Management Weaknesses Delayed Response to Flint Water Crisis Corrective Action Plan (CAP)

Supplemental EPA Response and OIG Disposition of Corrective Actions:

The EPA provided supplemental information on proposed corrective actions during a June 5, 2018, meeting with the OIG. Based on the revisions, the OIG accepted EPA responses as meeting the intent for eight of the nine recommendations. Eight recommendations are resolved pending completion of corrective actions. Recommendation 1 is unresolved.

Recommendation	Lead Office	CA	Target Date	Corrective Action
1. Establish controls to annually verify that the states are monitoring compliance with all Lead and Copper Rule requirements, including accurately identifying tier 1 sampling sites and maintaining continuous corrosion control.	OECA OW	1	PWSS Reviews: Beginning in 4th Quarter FY 2019 (September 30, 2019) and then ongoing. OECA national drinking water oversight approach recommendations: June 30, 2019.	EPA will continue to focus additional attention on metrics related to LCR implementation that can be incorporated into the protocol that Regions will use when conducting their annual Public Water System Supervision program reviews. The Annual Program Review template is being implemented in the FY 2018 and will continue to be adjusted for future reviews to meet oversight goals. For compliance monitoring and enforcement programs, OECA will continue to explore options for creating a national oversight approach for the drinking water programs and present recommendations to its senior managers by June 2019. Revised Corrective Action: OECA agrees on the value of national oversight for compliance monitoring and enforcement of the Drinking Water program, but it is vital that we evaluate and develop it in a collaborative manner with our State partners. Accordingly, we plan to work with the States to develop an approach or a pilot for implementing a national compliance monitoring and enforcement oversight program for the drinking water, as appropriate, by June 2019
2. Include in the revised Lead and Copper Rule the most protective protocols for monitoring and corrosion control. Revised recommendation: Revise the Lead and Copper Rule to improve the effectiveness of monitoring and corrosion control protocols.	OW	2	The estimated publication schedule for proposed revisions to the LCR is February 28, 2019.	The Agency is evaluating input we recently received from our state, local and tribal partners as well as the best available peer-reviewed science to ensure the Rule reflects the best ways to improve public health protection.

Recommendation	Lead Office	CA	Target Date	Corrective Action
3. Publicly document clear expectations, roles and responsibilities between the EPA and the state of Michigan in an official document, such as a memorandum of understanding or a supplemental primacy document.	Region 5	3	FY 2019 Annual Workplan which will be posted by October 1, 2018.	Region 5 documents clear expectations, roles and responsibilities between the EPA and each of our states in the annual workplan for the Public Water System Supervision (PWSS) program grant. To implement this recommendation, Region 5 will post the annual Michigan PWSS program workplan and end of year evaluation on EPA's website so that the information is publicly available. Revised Corrective Action: Region 5 will document clear expectations, roles and responsibilities between the EPA and each of our states in the annual workplan for the Public Water System Supervision (PWSS) program grant. To implement this recommendation, Region 5 will post the annual Michigan PWSS program workplan and end of year evaluation on EPA's website so that the information is publicly available.
4. Implement a system for regional drinking water staff, managers and senior leaders, which incentivizes staff elevating and management addressing important and emerging issues in accordance with the EPA's Policy on Elevation of Critical Environmental and Public Health Issues.	Region 5	4	PARS end-of- year for FY 2018, which must be completed by October 31, 2018.	Within Region 5, staff are encouraged to participate and share concerns in a variety of meetings including team, section, branch, division, and regional meetings. These regular check-in meetings among managers and staff allow for information to flow from regional leaders down to staff and vice versa. Region 5 agrees to explore ways to incentivize staff to elevate issues of concern during these engagement opportunities, including providing feedback during mid-year and end-of- year reviews with staff, specifically on the customer service critical element present in all staff PARS agreements. Region 5 will further explore ways to incentive staff and management elevating and addressing important and emerging issues including the use of current awards, both recognition and monetary. Revised Corrective Action: Within Region 5, staff are encouraged to participate and share concerns in a variety of meetings including team, section, branch, division, and regional meetings. These regular check-in meetings among managers and staff allow for information to flow from regional leaders down to staff and vice versa. Region 5 agrees to incentivize staff, managers, and senior leaders to elevate and address issues of concern during these engagement opportunities, including providing feedback during mid-year and end-of-year reviews, specifically on the customer service critical element present in all staff, managers, and senior leaders PARS agreements. Region 5 will incentive staff, management and senior leadership to elevate and address important and emerging issues including the use of current awards, both recognition and monetary

(Continued)

Recommendation	Lead Office	CA	Target Date	Corrective Action
5. Provide the public with all results from EPA reviews of Michigan's Safe Drinking Water Act program, and track the progress of identified corrective actions.	Region 5	5	Will post the corrective action table by September 30, 2018, and provide updates on progress by April 30 and September 30 [f]or each year thereafter.	The 2010 and 2016 Michigan Drinking Water Program Review Reports are currently available on-line. The FY 2010-2016 End-of-Year Evaluation Reports by Region 5 are available to the public through FOIA online. Region 5 and MDEQ are finalizing a Corrective Action Plan that has been developed from the recommendations in the 2016 Program Review Report, which Region 5 released in October of 2017. Once final and approved by Region 5, Region 5 will post the corrective action to the Agency's website and provide updates on a quarterly basis. In addition, the Michigan Corrective Action Plan will be attached to the FY 2018 and subsequent PWSS grant workplans so that both State and EPA commitments are clear. Frequent check-ins with the State are planned to ensure progress in implementing EPA's recommendations to strengthen Michigan's drinking water program.
6. Provide regular training for EPA drinking water staff, managers and leaders on Safe Drinking Water Act tools and authorities; state and agency roles and responsibilities; and any Safe Drinking Water Act amendments or Lead and Copper Rule revisions.	OECA OW	6	State/Regional staff LCR trainings conducted in all EPA Regions by May 31, 2018; SDWA Section 1431 training provided on May 23, 2018; Training efforts will be ongoing. EPA will reevaluate training needs by September 30, 2019.	EPA has provided and will continue to provide regular training nationally about SDWA tools and authorities, like Sections 1414 and 1431, and various NPDWRs, including LCR. As part of ongoing Agency efforts to enhance national implementation of the LCR, EPA has been providing training on the Rule's optimal corrosion control treatment and optimal water quality parameter requirements. The workshops provide a review of LCR requirements and emphasize the tools and authorities drinking water programs can leverage to implement the requirements more effectively. The training has been delivered through in-person workshops at each of the EPA Regions, as well as through special conference sessions. Revised Corrective Action: EPA has provided and will continue to provide regular training nationally to staff and managers about SDWA tools and authorities, like Sections 1414 and 1431, and various NPDWRs, including LCR. EPA will also make these trainings available to senior leaders.

Recommendation	Lead Office	CA	Target Date	Corrective Action
				As part of ongoing Agency efforts to enhance national implementation of the LCR, EPA has been providing training on the Rule's optimal corrosion control treatment and optimal water quality parameter requirements. The workshops provide a review of LCR requirements and emphasize the tools and authorities drinking water programs can leverage to implement the requirements more effectively. The training has been delivered through in-person workshops at each of the EPA Regions, as well as through special conference sessions
7. Implement a system to identify management risks in state drinking water programs, including elements such as atypical events, emerging public health concerns,environmental justice concerns and public health analysis.	OW	7	Initial findings and the workgroup's proposed framework for action are expected by December 31, 2018.	EPA has initiated a workgroup with participation from OECA, OW and the Regions. The workgroup will explore how best to use drinking water data and measures to identify public water systems that present or are likely to present a significant risk to public health. The workgroup will develop procedures and strategies to ensure timely and effective intervention where risks to public health are identified.
8. Create a system that tracks citizen complaints and gathers information on emerging issues. The system should assess the risk associated with the complaints, including efficient and effective resolution.	OW OECA	8	4th Quarter FY 2019; September 30, 2019.	Identify potential enhancements to existing systems and/or identify new system requirements that can support tracking of citizen complaints.
9. Improve oversight by establishing a clear and credible escalation policy for EPA intervention in states. The policy should provide steps the EPA will take when states do not act.	OECA OW	9	Training efforts will be ongoing. EPA will reevaluate training needs by September 30, 2019. Seek state input: June 30, 2018; Seek public comment: November 30, 2018; Decision on policy: July 31, 2019.	OECA will continue to provide training on SDWA authorities, including the use of SDWA Section 1431 when state and local authorities have not acted to protect public health. EPA has initiated a workgroup with participation from OECA, OW and the Regions. The workgroup will develop procedures and strategies to ensure timely and effective EPA intervention where a state's response to the risk is insufficient to protect the public's health.

APPENDIX D

Distribution

- The Administrator Deputy Administrator
- Special Advisor, Office of the Administrator Chief of Staff
- Agency Follow-Up Official (the CFO) Agency Follow-Up Coordinator General Counsel
- Assistant Administrator for Enforcement and Compliance Assurance Assistant Administrator for Water
- Regional Administrator, Region 5
- Associate Administrator for Congressional and Intergovernmental Relations Associate Administrator for Public Affairs
- Principal Deputy Assistant Administrator for Enforcement and Compliance Assurance Deputy Assistant Administrator for Enforcement and Compliance Assurance
- Deputy Assistant Administrator for Water
- Principal Deputy Assistant Administrator for Water Deputy Regional Administrator, Region 5
- Director, Office of Continuous Improvement, Office of the Administrator Director, Office of Children's Health Protection, Office of the Administrator
- Director, Office of Compliance, Office of Enforcement and Compliance Assurance
- Director, Office of Criminal Enforcement, Forensics and Training, Office of Enforcement and Compliance Assurance
- Director, Office of Ground Water and Drinking Water, Office of Water Director, Office of Water, Region 5
- Audit Follow-Up Coordinator, Office of the Administrator
- Audit Follow-Up Coordinator, Office of Enforcement and Compliance Assurance Audit Follow-Up Coordinator, Office of Water
- Audit Follow-Up Coordinator, Region 5 Public Affairs Coordinator, Region 5

In: Drinking Water
Editor: Erwin Jon Orosz

ISBN: 978-1-53617-284-3

Chapter 5

DRINKING WATER AND WASTEWATER INFRASTRUCTURE: OPPORTUNITIES EXIST TO ENHANCE FEDERAL AGENCY NEEDS ASSESSMENT AND COORDINATION ON TRIBAL PROJECTS*

United States Government Accountability Office

ABBREVIATIONS

EDA	Economic Development Administration
EPA	Environmental Protection Agency
HITS	Home Inventory Tracking System
HUD	Department of Housing and Urban Development
HIS	Indian Health Service
SDS	Sanitation Deficiency System
USDA	U.S. Department of Agriculture

WHY GAO DID THIS STUDY

Tens of thousands of American Indians and Alaska Natives do not have safe drinking water or wastewater disposal in their home—referred to as needs arising from a sanitation deficiency—at a higher percentage than the general population, according to IHS. Among other things, IHS assesses homes, either individually or by reviewing public water

* This is an edited, reformatted and augmented version of United States Government Accountability Office; Report to Congressional Requesters, Publication No. GAO-18-309, dated May 15, 2018.

systems, to determine any deficiencies. Seven agencies, including IHS, EPA, and USDA, have programs that provide drinking water and wastewater infrastructure assistance to Indian tribes.

GAO was asked to review federal efforts to provide water infrastructure assistance to Indian tribes. This report examines, among other objectives, the extent to which selected federal agencies (1) identified tribes' drinking water and wastewater infrastructure needs and (2) funded tribal water infrastructure projects, including tribes' most severe sanitation deficiencies.

GAO reviewed agency data on tribal needs, analyzed agency funding data for tribal water infrastructure projects, reviewed agency policy documents, and interviewed agency officials and officials from 22 tribes representing different geographic locations.

What GAO Recommends

GAO is making 16 recommendations, including that (1) IHS develop a way to indicate in its database if homes' deficiencies have been assessed and (2) IHS and USDA review and update project scoring processes. IHS agreed with these recommendations, and USDA proposed an approach for addressing the recommendation on scoring, as discussed in the report.

What GAO Found

Federal agencies have identified several billion dollars in existing and future tribal drinking water and wastewater infrastructure needs. Specifically, the Indian Health Service (IHS) worked with tribes to identify, in fiscal year 2016, an estimated $3.2 billion in water infrastructure projects to address existing sanitation deficiencies in Indian homes, and the Environmental Protection Agency (EPA) identified an additional $2.4 billion in future tribal drinking water infrastructure needs over the next 20 years. However, IHS could enhance the accuracy of its information about the water infrastructure needs of some Indian homes. In February 2018, the database that IHS uses to track Indian homes' sanitation deficiencies showed that about one-third of the homes (138,700) had no deficiency. However, because the database does not provide IHS with a way to record if a home's deficiency has been assessed, IHS could not determine whether these homes had no deficiency or if they had not yet been assessed to identify a deficiency. IHS officials stated that improving the database's accuracy would be beneficial. By implementing a way to indicate in its database whether these homes' deficiencies have been assessed, IHS could also more efficiently address any deficiencies in these homes.

Federal agencies provided about $370 million for tribal drinking water and wastewater infrastructure projects in fiscal year 2016, including some projects to address what the agencies identified as the most severe sanitation deficiencies (i.e., communities that lack safe drinking water or wastewater disposal). IHS and U.S. Department of Agriculture (USDA) policies direct the agencies to fund tribal projects that address these deficiencies. However, agency scoring processes may not always prioritize the projects that address them:

- IHS assigns points to projects using eight scoring factors, including sanitation deficiency and cost. Based on GAO's review of IHS documents and interviews with agency officials, IHS's process for selecting projects can discourage funding some projects that address the most severe sanitation deficiencies, especially those with a relatively high cost per home. As a result, some projects to serve homes without water infrastructure can remain unfunded for many years. IHS officials said the scoring factors balance a number of interests, and the agency is looking to improve the extent to which it funds projects that address these deficiencies.
- USDA uses a different set of scoring factors to assign points when evaluating project applications for its tribal water program, including rural population and income levels. However, USDA does not have a scoring factor to assign points to a project based on whether it will serve homes that lack safe drinking water or wastewater disposal, as it does with another program with similar goals. Instead, USDA officials said they use discretionary points to score projects on this basis, but these points may not be awarded at all. As a result, USDA may not have reasonable assurance that it consistently evaluates project applications in a way that aligns with agency policy to fund projects that address the most severe sanitation deficiencies.

By IHS reviewing and USDA updating their scoring processes, the agencies could have more assurance that the projects they fund address the most severe sanitation deficiencies in Indian communities.

May 15, 2018
Congressional Requesters

Tens of thousands of American Indians and Alaska Natives do not have safe, potable water available in their homes for drinking, cooking, and cleaning, or adequate facilities to safely dispose of wastewater, according to the Department of Health and Human Services' Indian Health Service (IHS). The agency has stated that, nationwide, members of Indian tribes are more likely to live in homes without safe drinking water and wastewater facilities than non-Indians in part due to the remoteness of many Indian reservations.[1] Moreover, IHS has reported that inadequate access to safe drinking water and wastewater disposal can lead to many health problems, including increased hospitalizations for pneumonia, influenza, and other illnesses.[2]

The federal government's role in funding construction and repair of drinking water and wastewater systems in Indian country differs from its role in funding such systems in the rest of the country.[3] According to IHS estimates, the federal government provides a significant share of the financial assistance to construct and repair drinking water and wastewater infrastructure in Indian country. In contrast, in non-tribal communities, utilities typically sell municipal bonds and raise revenues to pay for construction of drinking water and wastewater infrastructure by charging customers. These communities and utilities can receive funding and planning assistance from their respective states and the federal government to repair, replace, and upgrade their drinking water and wastewater infrastructure. In Indian country, tribes, tribally owned utilities, or separate entities are generally responsible for operating and maintaining drinking water and wastewater infrastructure. However, according to a national intertribal organization, tribes typically do not have access to the same array of financing options as state or local governments, and do not necessarily charge their customers for services.

Seven federal agencies administer programs that provide drinking water and wastewater infrastructure assistance to Indian tribes. IHS, the Environmental Protection Agency (EPA), and the U.S. Department of Agriculture (USDA) have drinking water and wastewater infrastructure programs that are specifically targeted to provide financial assistance for planning and construction to address Indian tribes' needs. According to IHS documentation, such needs arise from a sanitation deficiency in existing drinking water or wastewater infrastructure (or lack thereof) that can negatively affect public health. In addition, the Department of Housing and Urban Development (HUD), the Department of the Interior's Bureau of Reclamation, the U.S. Army Corps of Engineers, and the Department of Commerce's Economic Development Administration (EDA) administer programs that may provide financial assistance to tribes for drinking water and wastewater infrastructure. The types of assistance these agencies provide vary by program, and each program has its own eligibility requirements and authorities.

---

[1] This report focuses on federally recognized tribes. Federally recognized tribes have a government-to-government relationship with the United States and are eligible to receive certain protections, services, and benefits by virtue of their status as Indian tribes. The Secretary of the Interior is required by law to publish annually in the *Federal Register* a list of all Indian tribes that the Secretary recognizes as Indian tribes. As of January 29, 2018, there were 573 federally recognized tribal entities. Tribal members are individuals who are enrolled citizens or members of a tribe.

[2] Department of Health and Human Services, Indian Health Service, *Fiscal Year 2018 Justification of Estimates for Appropriations Committees* (Rockville, MD: May 23, 2017).

[3] Indian country is (1) all land within the limits of any Indian reservation, notwithstanding the issuance of any patent, and including rights-of-way running through the reservation; (2) all dependent Indian communities within the borders of the United States; and (3) all Indian allotments, the Indian titles to which have not been extinguished, including rights- of-way running through the same.

You asked us to review federal efforts to provide drinking water and wastewater assistance to Indian tribes, including interagency collaboration efforts. This report examines the extent to which the seven federal agencies, as applicable, (1) identified Indian tribes' drinking water and wastewater infrastructure needs; (2) funded tribal drinking water and wastewater infrastructure projects, including projects to address the most severe sanitation deficiencies;[4] and (3) collaborated to meet Indian tribes' drinking water and wastewater infrastructure needs.

To determine the extent to which these federal agencies identified Indian tribes' drinking water and wastewater infrastructure needs, we identified requirements for IHS and EPA to collect and report information on needs, but we did not identify such requirements for the other agencies. We reviewed IHS's and EPA's most recent reports describing tribal drinking water and wastewater infrastructure needs and reviewed the supporting data in IHS's Sanitation Deficiency System (SDS) on proposed tribal drinking water and wastewater infrastructure projects from fiscal year 2016, the most recent year of data available at the time of our review.[5] To assess the reliability of SDS project data and EPA needs information, we discussed the data and any of its limitations with agency officials and determined that the data were sufficiently reliable to provide descriptive information about tribes' drinking water and wastewater infrastructure project needs. We also assessed the reliability of specific information in IHS's Home Inventory Tracking System (HITS), including the sanitation deficiencies associated with homes, by reviewing documentation and interviewing IHS officials. As we discuss later in this report, we indicate when we found certain limitations with that information. We did not assess the reliability of other information in HITS that was not relevant to our review. We interviewed IHS and EPA officials involved with identifying tribal water needs from all 12 IHS areas and the 9 EPA regions that administer tribal drinking water and wastewater infrastructure programs.

We also interviewed officials from the other five agencies regarding any efforts to collect information on tribal drinking water and wastewater infrastructure needs.

To determine the extent to which the agencies funded tribal drinking water and wastewater infrastructure projects, we analyzed obligations data from the seven agencies that administer programs that provide drinking water and wastewater infrastructure assistance to Indian tribes— IHS, EPA, USDA, HUD, Reclamation, Corps, and EDA— from fiscal years 2012 through 2016.[6] To assess the reliability of the data obtained from

[4] IHS refers to these deficiencies as tribes' greatest needs—that is, homes and communities that lack safe drinking water or wastewater disposal, or both. We use the term most severe sanitation deficiencies to describe these deficiencies in this report.

[5] In conducting this work, we relied on the concept of needs as defined by IHS and EPA in their reports as opposed to independently defining the concept of need or evaluating the legitimacy of the reported needs. According to IHS, needs arise from a sanitation deficiency in existing drinking water or wastewater infrastructure (or lack thereof) that can negatively affect public health. According to EPA, its estimates of needs represent infrastructure projects necessary for water systems to continue to provide safe drinking water to the public.

[6] We are using the term funding to refer to obligations. Obligations are definite commitments that create a legal liability of the government for the payment of goods and services ordered or received, or a legal duty on the

these agencies, we discussed the funding data, each agency's data system, and any limitations with agency officials and determined that the data were sufficiently reliable for the purposes of our reporting objective. To determine the extent to which agency funding addressed the most severe sanitation deficiencies, we identified programs that have documented goals in regulation and policy to fund projects that address these needs.[7] For these programs, we compared the number of funded projects that addressed the most severe deficiencies with the number of funded projects that met other needs for fiscal year 2016. During the course of reviewing these regulations and agencies' obligations data, we identified issues with how USDA awarded grants under one of its tribal programs—the Rural Alaska Village Grant program.[8] We then obtained and reviewed documentation of these grant awards and interviewed officials who manage the program and officials from the agency's Office of the General Counsel.

To determine the extent to which the federal agencies collaborated to meet tribal water needs, we reviewed documentation of national-level interagency collaboration, including memorandums of understanding. We compared the agencies' actions to the key features of interagency collaboration that we have previously identified.[9]We reviewed agencies' collaboration at the regional level by surveying the seven agencies about their joint actions on activities related to tribal drinking water and wastewater in six states—Alaska, Arizona, California, New York, Oklahoma, and South Dakota—and by conducting a network analysis using the survey responses. We selected regional offices that operate in individual states as the unit of analysis because the federal agencies organize their field structures differently, with some using region, area, and state offices to work with tribes—we refer to all of these office types as regional offices. We selected this nonprobability sample of six states to include a large percentage of the number of federally recognized tribes, to obtain a range in the total federal obligations to tribes and the identified needs of tribes, and for geographic diversity. The sample of states is not generalizable, and the results of our work do not apply to all states where Indian tribes are located. However, they provide illustrative examples of interagency collaboration

---

part of the United States that could mature into a legal liability by virtue of actions on the part of the other party beyond the control of the United States. The Corps did not provide obligations data for the requested time period because the agency did not fund tribal water infrastructure projects during these years.

[7] IHS's Sanitation Facilities Construction program, EPA's clean water set-aside program, and USDA's Native American program identify the most severe sanitation deficiencies as the absence of safe drinking water or wastewater disposal facilities, or both. EPA's drinking water set-aside program identifies these deficiencies similarly, but we did not include this program in our analysis because of the variation in how EPA regions implement the program.

[8] The Rural Alaska Village Grant program is authorized by 7 U.S.C. § 1926d, and is also known as the section 306D grant program because it is authorized by section 306D of the Consolidated Farm and Rural Development Act.

[9] GAO, *Managing for Results: Key Considerations for Implementing Interagency Collaborative Mechanisms*, GAO-12-1022 (Washington, D.C.: Sept. 27, 2012). Key features fall into the following categories: outcomes and accountability, bridging organizational cultures, leadership, clarity of roles and responsibilities, participants, resources, and written guidance and agreements.

within the six selected states, which include about 70 percent of the 573 federally recognized tribes.

We sent collaboration surveys to the regional offices of each of the seven federal agencies in the six states. We also sent a survey to the State of Alaska because the state provides a 25 percent match for two federal water infrastructure programs, but we did not include other state agencies in our survey because they do not provide a similar match. The survey asked each agency regional office in each state about the mechanisms that it used to collaborate with each of the other regional offices within the same state for the 3-year period prior to our survey, the factors that affected their collaboration, and any additional mechanisms that would be beneficial to use in the future. We disseminated a total of 46 surveys and obtained and analyzed 46 responses. We conducted site visits to three of the six states, selected based on their geographic diversity and the range of identified water infrastructure needs. We also interviewed federal agency officials and officials with the State of Alaska about their collaboration in the six states. We interviewed officials from 22 Indian tribes and representatives from 8 intertribal organizations to obtain their perspectives on obtaining drinking water and wastewater infrastructure assistance from the seven agencies in the six selected states. We selected these tribes to obtain a range in their geographic locations and the amount of drinking water and wastewater infrastructure assistance they have received. Our findings are not generalizable to all tribes but provide illustrative examples. Appendix I contains a detailed description of our scope and methodology, and appendix II contains additional information about our survey and network analysis, including questions and summary responses.

We conducted this performance audit from August 2016 to May 2018 in accordance with generally accepted government auditing standards. Those standards require that we plan and perform the audit to obtain sufficient, appropriate evidence to provide a reasonable basis for our findings and conclusions based on our audit objectives. We believe that the evidence obtained provides a reasonable basis for our findings and conclusions based on our audit objectives.

BACKGROUND

This section provides information on water infrastructure in Indian country, federal programs that provide drinking water and wastewater infrastructure assistance to Indian tribes, and our prior work on interagency collaboration.

Water Infrastructure in Indian Country

The 573 federally recognized Indian tribes in the United States vary greatly in terms of their culture, language, population size, land base, location, and economic status. Many are located in remote and often environmentally challenging areas. According to the U.S. Census Bureau's American Community Survey,[10] in 2016, about 26 percent of American Indians and Alaska Natives were living below the poverty line, compared with 14 percent for the nation as a whole.[11]

According to EPA databases, tribes operate about 950 public drinking water systems and about 340 public wastewater systems.[12] Drinking water systems often include groundwater wells, water treatment plants, and pipelines to deliver water to homes. A regulated, centralized wastewater system may include sewer lines, tanks, and wastewater treatment plants or lagoons, but small, rural communities are more likely to have decentralized wastewater systems, such as individual septic systems. Once centralized water or wastewater systems are constructed in Indian country, ownership is typically transferred to the tribe. A tribally owned utility, tribal government, or a separate entity operates and maintains the system on behalf of the tribe. Some tribal utilities charge user fees to help offset operations and maintenance costs, but other tribal utilities do not charge these fees because of users' low income levels or for cultural reasons, according to IHS and tribal officials.

According to EPA, thousands of Indian homes are not currently served by a regulated, centralized drinking water or wastewater system, due in part to the logistical and other challenges associated with Indian water systems that must serve widely dispersed populations in remote locations.[13] Instead, as we reported in September 2017, homes that are not served by water systems may have private wells and septic systems,[14] or they may be entirely unserved. According to EPA and IHS documents, some tribal

---

[10] The U.S. Census Bureau's American Community Survey continuously collects data on social, demographic, economic, and housing characteristics that help determine how federal funds are allocated to states and localities and that provide information to communities to aid in planning investments and services. The Bureau collects this data on a monthly sample of households and aggregates the results into 1-, 3-, and 5-year estimates, depending on the population size of the area.

[11] U.S. Census Bureau, *2016 American Community Survey 1-Year Estimates, Selected Population Profiles in the United States*, accessed December 5, 2017. Results include people who identified as American Indian or Alaska Native alone and not in combination with another race, compared with the total population.

[12] A public water system under the Safe Drinking Water Act is a system for the provision to the public of water for human consumption through pipes or other constructed conveyances that has at least 15 service connections or regularly serves at least 25 individuals. 42 U.S.C. § 300f(4)(A). According to EPA, the agency regulates the tribally owned public drinking water systems for every tribe except the Navajo Nation, which has assumed regulation of its public drinking water systems. Also, any individual tribe may have multiple public water systems. For wastewater systems, according to EPA officials, the approximately 340 public wastewater systems operated by tribes are those systems with permits to discharge pollutants into waters of the United States.

[13] Environmental Protection Agency, *Drinking Water Infrastructure Needs Survey and Assessment: Fifth Report to Congress*, EPA 816-R-13-006 (Washington, D.C.: April 2013).

[14] GAO, *Drinking Water and Wastewater Infrastructure: Information on Identified Needs, Planning for Future Conditions, and Coordination of Project Funding*, GAO-17-559 (Washington, D.C.: Sept. 20, 2017).

members may haul drinking water from a regulated drinking water source. However, containers used to haul and store the water can introduce bacteria and other contaminants. Also, because the regulated water source in some communities may be many miles away, residents may haul drinking water from nearby unregulated water sources, such as streams or livestock wells. For homes without access to a wastewater disposal system, residents may use a privy, use honeybuckets, or discharge waste directly to the ground.[15]

According to researchers with the Centers for Disease Control and Prevention, restricted access to clean water for hand washing and hygiene, along with manually disposing of waste, exposes people— especially infants and the elderly—to higher rates of illness and hospitalization.[16] We reported in January 2017 that such health concerns underscore the importance of quality health care—including preventative care, such as providing safe sanitation facilities—for American Indian and Alaska Native people.[17] Further, according to IHS, American Indian and Alaska Native families living in homes with satisfactory environmental conditions, which include safe water and sewer systems, require appreciably fewer medical services and place fewer demands on primary health care delivery systems.[18]

Federal Drinking Water and Wastewater Infrastructure Programs to Assist Indian Tribes

Seven federal agencies administer a number of programs that provide assistance to tribes for drinking water and wastewater infrastructure projects. Each agency has its own programs and processes for providing this assistance, with some similarities. Tribes can apply to one or more federal programs for financial assistance. In some cases, federal agencies coordinate to jointly fund the same project if the project is too large for one agency to fund. In other cases, agencies may work together by separately funding different parts of a large project or different phases of a multi-year project. Of these agencies, IHS, EPA, and USDA administer drinking water and wastewater infrastructure programs that are specific to Indian tribes.

---

[15] A honeybucket is a bucket used as a toilet that does not use water and has to be emptied manually.

[16] See, for example, Thomas W. Hennessy, et al., "The Relationship between In-Home Water Service and the Risk of Respiratory Tract, Skin, and Gastrointestinal Tract Infections Among Rural Alaska Natives," *American Journal of Public Health,* vol. 98,no. 11 (2008); and Thomas W. Hennessy and Jonathan M. Bressler, "Improving Health in the Arctic Region Through Safe and Affordable Access to Household Running Water and Sewer Services: an Arctic Council Initiative," *International Journal of Circumpolar Health,* vol. 75, no. 1 (2016).

[17] GAO, *Indian Health Service: Actions Needed to Improve Oversight of Quality of Care,* GAO-17-181 (Washington, D.C.: Jan. 9, 2017).

[18] We initially reported on this more than 40 years ago. See GAO, *Progress and Problems in Providing Health Services to Indians*, B-164031(2) (Washington, D.C.: Mar. 11, 1974).

HIS

IHS's mission is to raise the physical, mental, social, and spiritual health of American Indians and Alaska Natives to the highest level. To fulfill this mission, IHS provides primary health care and disease prevention services.[19] IHS's Office of Environmental Health and Engineering's Sanitation Facilities Construction program, established in 1959, contributes to IHS's disease prevention efforts.[20] This program provides technical and financial assistance to Indian tribes for the cooperative development and construction of drinking water and wastewater systems and support facilities. According to the Indian Health Care Amendments of 1988, it is the policy of the United States that all Indian communities and Indian homes, new and existing, be provided with safe and adequate water supply systems and sanitary wastewater disposal systems as soon as possible.[21] IHS's 12 regional offices, called Areas, are responsible for working with tribes when administering the Sanitation Facilities Construction program.

The Indian Health Care Amendments of 1988 require that IHS report annually to Congress on the sanitation deficiency levels for Indian tribes and communities, including, among other things, the amount of funds necessary to raise all Indian tribes and communities to zero sanitation deficiency.[22] The act identifies five deficiency levels, and IHS uses a deficiency level of 0 to represent the absence of a deficiency in its data systems (see Table 1).[23]

To develop its annual report to Congress and identify sanitation deficiencies in Indian communities and homes, IHS maintains two data systems: (1) the Sanitation Deficiency System (SDS), which contains proposed drinking water and wastewater infrastructure projects to address identified sanitation deficiencies; and (2) the Home Inventory Tracking System (HITS), which contains home-specific information that complements the SDS's project-specific information.[24] According to IHS program documentation, the

---

[19] In February 2017, we added improving federal management of Indian programs—Indian health care, Indian education, and Indian energy—to our list of areas at high risk due to greater vulnerabilities to fraud, waste, abuse, and mismanagement or due to a need for broad-based transformation. We added this area in part because we found numerous challenges that IHS faces in administering health care services, including inadequate oversight of health care that hindered IHS's ability to ensure quality care in Indian communities. See GAO, *High Risk Series: Progress on Many High-Risk Areas, While Substantial Efforts Needed on Others*, GAO-17-317 (Washington, D.C.: Feb. 15, 2017). IHS's Sanitation Facilities Construction program was not included in the scope of the high risk designation.

[20] Pub. L. No. 86-121, 73 Stat. 267 (July 31, 1959) (*codified at* 42 U.S.C. § 2004a). The Sanitation Facilities Construction program also works with tribes to identify and address solid waste sanitation deficiencies. We do not include solid waste deficiencies in this report.

[21] 25 U.S.C. § 1632(a)(5). IHS policy states that an Indian community is one in which more than 50 percent of residents are American Indian or Alaska Native. The policy also states that an Indian home is a home owned by an American Indian or Alaska Native and can include a home owned by a tribe that is leased to an American Indian or Alaska Native in certain circumstances. Homes eligible to receive Sanitation Facilities Construction program assistance are 24-hour, year-round family dwellings and are located in a Purchased/Referred Care Delivery Area.

[22] 25 U.S.C. § 1632(g).

[23] According to IHS documentation, a sanitation deficiency is a need arising from existing drinking water or wastewater infrastructure (or lack thereof) that can negatively affect public health.

[24] SDS and HITS are two components of IHS's Sanitation Tracking and Reporting System. This system also includes components that track ongoing projects and new homes needing assistance, among others.

project descriptions in the SDS are to include information about the sanitation deficiency level that each project will address, the project's estimated cost, and the number of Indian homes that the project will serve.[25] According to IHS documents, HITS is to include information about each Indian home that may have a sanitation deficiency that is eligible to receive Sanitation Facilities Construction program assistance, including the home's geographic location and deficiency level. Eligible homes can be located on or off reservations, but according to IHS officials, the agency typically does not collect information about Indian homes located in large urban areas. According to IHS program documentation, IHS uses information in HITS to track the status of and plan for the provision of sanitation facilities for Indian homes.

Table 1. Drinking water and wastewater sanitation deficiency levels used by the Indian Health Service (IHS)

Deficiency level	Description of sanitation deficiency
5	Community or home that lacks a safe water supply and a sewage disposal system.
4	Community or home that lacks either a safe water supply system or a sewage disposal system.
3	Community or home that has an inadequate or partial water supply and a sewage disposal facility that does not comply with applicable water supply and pollution control laws.[a]
2	Sanitation system that complies with all applicable water supply and pollution control laws, and in which the deficiencies relate to capital improvements that are necessary to improve the facilities to meet the needs of the tribe or community for domestic sanitation facilities.
1	Sanitation system that complies with all applicable water supply and pollution control laws, and in which the deficiencies relate to routine replacement, repair, or maintenance needs.
0	Sanitation facilities are adequate.

Source: GAO analysis of IHS information. | GAO-18-309.

Note: The Indian Health Care Amendments of 1988 define deficiency levels 1 through 5. IHS uses deficiency level 0 to indicate the absence of a deficiency in its data systems.

[a]IHS also uses deficiency level 3 for a community or home that does not have a solid waste disposal facility.

To address tribes' identified sanitation facility needs, IHS is authorized to construct essential sanitation facilities, including domestic and community water supplies and facilities, as well as wastewater disposal facilities for Indian homes, communities, and lands.[26] Under the Sanitation Facilities Construction program, IHS administers two primary drinking water and wastewater infrastructure activities: one to address sanitation deficiencies in existing homes and communities based on needs identified in the SDS, and one to provide water infrastructure for newly constructed or recently renovated Indian homes—these needs are not included in the SDS.[27] According to IHS policy, the

[25] According to IHS policy, when eligible Indian homes are located within a non-Indian community with a population of less than 10,000, and a sanitation deficiency exists in the homes, IHS will include the prorated cost to correct the deficiency for the Indian homes as a project in the SDS.

[26] 42 U.S.C. § 2004a(a)(1).

[27] The Sanitation Facilities Construction program also funds two other types of water infrastructure projects: special projects, such as research studies, and emergency projects. The program, however, does not provide water infrastructure for non-Indian homes and non-residential users, such as commercial entities. IHS also does not

agency selects projects to fund that address deficiencies in existing homes based on ranked project lists contained in the SDS, by area.

According to IHS policy, the agency can manage sanitation projects on behalf of a tribe (direct service), or a tribe or tribal entity can elect to manage projects.[28] According to this policy, to implement a project under direct service, a tribe formally requests IHS assistance, and IHS engineers typically develop projects to include in the SDS. When IHS selects the project to fund, the tribe decides whether it will complete the project design and manage the construction contract or have IHS engineers do so.

EPA

EPA provides annual grants to states to help finance drinking water and wastewater infrastructure through its Drinking Water and Clean Water State Revolving Fund programs, respectively.[29] EPA sets aside a certain percentage of the appropriations it receives for these programs to make grants directly to Indian tribes for drinking water and wastewater infrastructure.[30] Nine EPA regions administer the Drinking Water Infrastructure Grants Tribal Set-Aside and the Clean Water Indian Set- Aside programs,[31] while states administer the State Revolving Funds.

Under the drinking water set-aside program, EPA funds projects for community water systems that serve tribal populations, as well as for non- profit, non-community water systems owned by a tribal government that serve a tribal population. Under the clean water set-aside program, EPA provides funding for the planning, design, and construction of wastewater treatment plant facilities that serve federally recognized Indian tribes, Alaska Native villages, and certain tribes in Oklahoma. According to EPA officials, tribes are among those eligible to receive loans from the states' State Revolving Fund programs. In addition, EPA administers the separate Alaska Native Villages and

---

provide financial assistance for routine operation and maintenance costs, such as paying a drinking water or wastewater system operator's salary.

[28] The Indian Self-Determination and Education Assistance Act, as amended, authorizes Indian tribes to assume responsibility for administration of certain federal programs from IHS and Interior through a self-determination contract or self-governance compact. Pub. L. No. 93-638, 88 Stat. 2203 (Jan. 4, 1975) (*codified as amended at* 25 U.S.C. §§ 5301-5423). As of November 2017, 19 federally recognized tribes in the continental United States administer their own sanitation facility construction programs, and the Alaska Native Tribal Health Consortium administers a sanitation facility construction program for the more than 200 federally recognized tribes in Alaska.

[29] States use these grants, and provide a required minimum 20 percent match, to capitalize their revolving funds. Pub. L. No. 100-4, § 212, 101 Stat. 7, 21-28 (Feb. 4, 1987) (*codified as amended at* 33 U.S.C. §§ 1381-1388) (Clean Water State Revolving Funds); Pub. L. No. 104-182, § 130, 110 Stat. 1613, 1662–1672 (Aug. 4, 1996) (*codifiedas amended at* 42 U.S.C. § 300j-12) (Drinking Water State Revolving Funds).

[30] The Clean Water Act as amended requires EPA to set aside a percentageof the amount appropriated for, among other things, state water pollution control revolving funds for grants for projects and activities to serve Indian tribes; this is known as the Clean Water Indian Set-Aside program. 33 U.S.C. § 1377(c). The Safe Drinking Water Act as amended authorizes EPA to set aside a percentage of the amount appropriated to carry out the state drinking water revolving loan funds for grants to Indian tribes; this is known as the Drinking Water Infrastructure Grants Tribal Set-Aside program.42 U.S.C. § 300j-12(i)(1).

[31] According to EPA, Region 3 is the only region that does not administer the clean water or drinking water set-aside programs.

Rural Community Water Grant program that awards grants to the State of Alaska to, among other things, improve sanitation in rural and Alaska Native villages.[32]

USDA

USDA's Rural Utilities Service allocates a portion of its appropriation for rural water and wastewater disposal programs to make drinking water and wastewater infrastructure grants to Indian tribes; this is referred to as the Native American program.[33] USDA administers the Native American program at the national level and works with tribes at the state office and local level to conduct outreach and assist with the application process, among other things. The Native American program provides grants for water and wastewater facilities and services to rural and low-income tribal communities "whose residents face significant health risks ... due to the fact that a significant proportion of the community's residents do not have access to, or are not served by, adequate affordable water supply systems or waste disposal facilities."[34] In addition, USDA administers the Rural Alaska Village Grant program, which provides grants to the State of Alaska for development and construction of water and wastewater systems that address dire sanitation conditions in rural or Alaska Native villages with 10,000 or fewer people.[35] Tribes are also eligible to receive loans and grants for infrastructure investments from the agency's Water and Waste Disposal Program, which is administered by USDA's state offices.[36] Tribes that are located close to the U.S.-Mexico border and that meet the definition of a colonia are eligible for assistance from USDA's Colonias program, a water infrastructure grant program to serve state- designated, low-income, unincorporated areas along the border.[37] Finally, USDA administers a grant program to provide technical assistance and training,[38] and the agency makes pre-planning grants available to tribes,

[32] The Alaska Native Villages and Rural Community Water Grant program is authorized by 33 U.S.C. § 1263a.

[33] The Native American program is authorized by 7 U.S.C. § 1926c, which is also known as section 306C of the Consolidated Farm and Rural Development Act.

[34] 7 U.S.C. § 1926c (a)(1). A facility is eligible for section 306C assistance if it provides water or waste disposal services, or both, in rural areas where the per capita income of residents in the county is not more than 70 percent of the most recent national average per capita income, as determined by the Department of Commerce, and the unemployment rate of the residents is not less than 125 percent of the national average unemployment rate, as determined by the Bureau of Labor Statistics. These eligibility requirements do not apply to colonias—any identifiable community designated in writing by the state or county it is located in, determined to be a colonia on the basis of objective criteria, and that existed and was generally recognized as a colonia before October 1, 1989.

[35] The Rural Alaska Village Grant program is authorized by 7 U.S.C. § 1926d.

[36] The Water and Waste Disposal program is authorized by 7 U.S.C. § 1926, otherwise known as section 306 of the Consolidated Farm and Rural Development Act.

[37] The Colonias program is authorized by 7 U.S.C. § 1926c. Colonias are communities of all types and sizes, both incorporated and unincorporated, and can be, but are not usually, within tribal reservations. GAO, *Rural Water Infrastructure: Improved Coordination and Funding Processes Could Enhance Federal Efforts to Meet Needs in the U.S.-Mexico Border Region,* GAO-10-126 (Washington, D.C.: Dec. 18, 2009).

[38] USDA's Technical Assistance and Training program is authorized by section 306(a)(14) of the Consolidated Farm and Rural Development Act. The program awards grants to private non-profit organizations to enable them to provide technical assistance and training to Indian tribes and other entities to, among other things, improve the operation and maintenance practices at existing water and wastewater works in rural areas. 7 U.S.C. § 1926(a)(14).

organizations that serve tribes, and other recipients through multiple programs to assist with the development of application components, such as preliminary engineering or environmental reports.[39]

Additional Agencies

Four additional agencies may provide drinking water or wastewater assistance to Indian tribes through other programs not specific to drinking water or wastewater or as authorized by statute:

- HUD. HUD administers the Indian Community Development Block Grant program, a set-aside from the agency's Community Development Block Grant program that is specific to Indian tribes.[40] Indian Community Development Block grants can be used for construction of public facilities, provision of public services, housing, and certain economic development projects, among other things. HUD also awards Indian Housing Block Grants to tribes for affordable housing activities, which may include the development and rehabilitation of utilities, necessary infrastructure, and utility services.[41]
- Reclamation. As authorized by statute, Reclamation provides assistance for drinking water infrastructure in the 17 western states, including rural water supply projects for tribes. Some of the statutes that direct Reclamation to construct rural water supply projects for tribes are enacted Indian water rights settlements. In addition, until September 2016, Reclamation's rural water supply program was authorized to conduct appraisal investigations and feasibility studies for proposed rural water supply projects, including those that serve Indian tribes, but the program was not authorized to construct rural water supply projects.[42]
- Corps. As authorized by statute, the Corps may provide designated communities, counties, and states with design and construction assistance for drinking water and wastewater infrastructure. For example, Congress has authorized and made appropriations for the Corps to provide assistance to Indian tribes for water-related environmental infrastructure projects—including wastewater treatment

---

[39] Programs that offer pre-planning grants include the Rural Alaska Village Grant, Water & Waste Disposal Predevelopment Planning Grants, and Special Evaluation Assistance for Rural Communities and Households.

[40] The Indian Community Development Block Grant program is authorized by 42 U.S.C. § 5306(a)(1) and is administered by the regional offices of HUD's Office of Native American Programs.

[41] Other eligible activities include development of affordable housing, provision of housing- related services for affordable housing, and crime prevention and safety activities. The Indian Housing Block Grant is authorized by the Native American Housing Assistance and Self-Determination Act of 1996, Pub. L. No. 104-330, 110 Stat. 4016 (Oct. 26, 1996) (*generally codified as amended at* 25 U.S.C. §§ 4101-4243).

[42] The Rural Water Supply Act of 2006 authorized Reclamation to establish a rural water supply program in western states. Pub. L. No. 109-451, tit. I., 120 Stat. 3345 (Dec. 22, 2006). The authority to carry out the program expired at the end of fiscal year 2016.

facilities and water supply, storage, treatment, and distribution facilities—through the Corps' Section 219 Environmental Infrastructure Program.[43]

- EDA. EDA's Public Works Program provides grants to economically distressed areas to, among other things, help rehabilitate, expand, and improve their public works facilities, including drinking water and wastewater facilities.[44] The Economic Adjustment Assistance Program provides grants for, among other things, development of public facilities, including drinking water and wastewater facilities.[45] EDA's Planning Program provides grants to various entities, including tribes, to pay the costs of economic development planning, which can include planning for water infrastructure.[46]

Prior GAO Work on Interagency Collaboration

As part of our body of work on interagency collaboration, our September 2012 report discussed a variety of mechanisms to implement interagency collaborative efforts and identified key features that all efforts benefit from.[47] Mechanisms to implement interagency collaborative efforts include establishing interagency task forces or signing memorandums of understanding. Key features, many of which are related to practices to enhance and sustain collaboration identified in our previous work, fall into the following categories: outcomes and accountability, bridging organizational cultures, leadership, clarity of roles and responsibilities, participants, resources, and written guidance and agreements.[48]

---

[43] The Corps is authorized to provide assistance to nonfederal entities for specified water- related infrastructure projects, including wastewater treatment and related facilities and water supply, storage, treatment, and distribution facilities, by section 219 of the Water Resources and Development Act of 1992 as amended. Pub. L. No. 102-580, § 219, 106 Stat. 4797, 4835-4836 (Oct. 31, 1992).

[44] EDA's Public Works Program is authorized by section 201 of the Public Works and Economic Development Act of 1965, Pub. L. No. 105-393, tit. I, § 102(a), 112 Stat. 3601, 3601 (Nov. 13, 1998) (*codified as amended at* 42 U.S.C. § 3141).

[45] EDA's Economic Adjustment Assistance Program is authorized by section 209 of the Public Works and Economic Development Act of 1965, Pub. L. No. 105-393, tit. I, § 102(a), 112 Stat. 3601, 3605 (Nov. 13, 1998) (*codified as amended at* 42 U.S.C. § 3149).

[46] EDA's Planning Program is authorized by section 203 of the Public Works and Economic Development Act of 1965, Pub. L. No. 105-393, tit. I, § 102(a), 112 Stat. 3601, 3602 (Nov. 13, 1998) (*codified as amended at* 42 U.S.C. § 3143).

[47] GAO-12-1022.

[48] In October 2005, we described how agencies can enhance and sustain their interagency collaborative efforts by engaging in eight practices, including: defining and articulating a common outcome; establishing mutually reinforcing or joint strategies; identifying and addressing needs by leveraging resources; and establishing compatible policies, procedures, and other means to operate across agency boundaries. See GAO, *Results-Oriented Government: Practices That Can Help Enhance and Sustain Collaboration among Federal Agencies,* GAO-06-15 (Washington, D.C.: Oct. 21, 2005).

FEDERAL AGENCIES ESTIMATED ABOUT $3 BILLION IN EXISTING TRIBAL DRINKING WATER AND WASTEWATER INFRASTRUCTURE NEEDS IN FISCAL YEAR 2016, BUT THE NEEDS ARE UNDERESTIMATED

IHS and EPA estimated costs for tribal water infrastructure needs, with IHS identifying at least $3.2 billion in estimated costs for infrastructure projects to address existing drinking water and wastewater infrastructure needs for fiscal year 2016 and EPA estimating the costs of future tribal drinking water infrastructure needs at an additional $2.4 billion over the following 20 years. However, IHS's estimate of existing needs is likely too low because IHS has not identified all eligible Indian homes that may have existing sanitation deficiencies—drinking water or wastewater infrastructure needs—and some data in the system that IHS uses to track home-specific infrastructure needs are not accurate.

IHS and EPA Have Estimated Several Billion Dollars in Existing and Future Tribal Water Infrastructure Needs

In fiscal year 2016, IHS identified about $3.2 billion in estimated costs for projects to address existing tribal drinking water and wastewater infrastructure needs. This estimate represented more than 2,000 projects in the SDS to address 373 tribes' existing drinking water and wastewater infrastructure needs.[49] To develop these projects, IHS policy directs area staff to invite all federally recognized tribes to identify existing drinking water and wastewater infrastructure needs each year.[50] IHS staff then work with interested tribes to develop projects, including cost estimates, to include in the SDS. In fiscal year 2016, projects to address deficiency levels 4 and 5—homes or communities that lack a safe drinking water supply or wastewater disposal system, or both—accounted for $1.6 billion, or about half, of the total estimated costs of tribal infrastructure needs in the SDS. More than 80 percent of the deficiency level 4 and 5 project costs were located in the IHS Alaska and Navajo areas.[51] In addition, in fiscal year 2016, IHS determined that more than 60 percent of the total existing drinking water and wastewater infrastructure needs in

[49] According to IHS officials, they have not identified the number of unique homes to be served by these projects because one home can be served by multiple projects.

[50] IHS is required by statute to consult with Indian tribes and tribal organizations to determine the sanitation needs of each tribe when preparing the agency's annual reports to Congress.

[51] IHS's Alaska Area provides assistance to Alaska Natives in the state of Alaska, and the Navajo Area provides assistance to American Indians in or near the Navajo Nation.

the SDS were infeasible,[52] mostly due to the significant costs associated with infeasible deficiency level 5 projects.[53]

EPA collects and reports data on the drinking water infrastructure needs of the nation's public water systems, including the future needs of tribally owned or operated drinking water systems. Specifically, EPA is required to assess capital improvement needs of all eligible public water systems every 4 years, and EPA has conducted its Drinking Water Infrastructure Needs Survey and Assessment to obtain this information every 4 years from 1995 through 2015.[54] EPA last reported in 2013 on the estimated costs of capital improvement projects needed to repair, replace, and upgrade existing tribal and other public drinking water systems over the following 20 years.[55] In its 2013 report, EPA estimated the costs of future tribal drinking water needs of public systems at approximately $2.4 billion.[56] EPA does not, and is not required to collect information about future tribal wastewater infrastructure needs.[57]

Other agencies that provide tribes with assistance for drinking water or wastewater infrastructure projects do not—and are not required to— systematically identify tribal drinking water or wastewater infrastructure needs. For example, USDA officials explained that tribes identify needs through the applications they submit to the agency's programs. These officials stated that they also identify tribal needs through outreach to tribes and coordination with other agencies, such as IHS. In addition, HUD officials said that they do not collect information specifically about tribal water infrastructure needs because they rely on the tribes to propose or identify projects to meet any needs based on the tribes' priorities.[58]

---

[52] For all projects in the SDS, IHS determines whether the project is considered economically feasible or infeasible, based on whether project costs fall below established maximum allowable per-home construction costs.

[53] IHS estimated in fiscal year 2016 that infeasible deficiency level 5 projects would cost more than $1 billion.

[54] The Safe Drinking Water Act Amendments of 1996 require EPA to conduct an assessment of water system capital improvement needs of all the nation's eligible public water systems and report to Congress on the results every 4 years. 42 U.S.C. § 300j- 12(h). In accordance with a schedule that is consistent with this needs assessment, the act also requires EPA to prepare surveys and assess the needs of drinking water treatment facilities to serve Indian tribes. 42 U.S.C. § 300j-12(i)(4). EPA conducted a survey in 2015, but the agency had not released the report with its findings at the time of our review.

[55] Environmental Protection Agency, *Drinking Water Infrastructure Needs Survey and Assessment: Fifth Report to Congress*, EPA 816-R-13-006 (Washington, D.C.: April 2013).

[56] In its 2013 report, EPA estimated that overall tribal drinking water needs totaled about $3.3 billion. However, the report stated that approximately 27 percent of the estimated tribal needs were also associated with IHS's estimate of existing needs in the SDS. Therefore, $2.4 billion represents the future drinking water infrastructure needs reported by EPA minus the existing needs IHS included in the SDS's estimate of needs. EPA's estimate does not include needs associated with private infrastructure, such as future costs to repair or replace private drinking water wells, and it does not include needs associated with projects related primarily to population growth.

[57] The Clean Water Act, as amended, requires EPA to biennially make a detailed estimate of the cost of construction of all needed publicly owned treatment works in the country and submit it to Congress. 33 U.S.C. § 1375(b)(1). In its January 2016 submittal, EPA stated that it did not include tribal wastewater infrastructure in its Clean Watersheds Needs Survey since the agency uses IHS's SDS to identify any wastewater infrastructure projects on tribal lands.

[58] In January 2017, HUD issued a report on Indian housing needs that found that 5.6 percent of tribal homes have a plumbing inadequacy—defined as lacking piped hot water or a flush toilet, or lacking both bathtub and shower, for the exclusive use of the unit—compared to 1.3 percent of all homes in the United States. Department of Housing and Urban Development, *Housing Needs of American Indians and Alaska Natives in*

IHS Underestimates Existing Tribal Water Infrastructure Needs Because IHS Has Not Identified All Eligible Homes with Infrastructure Needs and Relevant Data Are Not Accurate

IHS area staff work with tribes each year to (1) identify Indian homes eligible for and in need of IHS drinking water or wastewater infrastructure assistance to include in IHS's home-specific tracking system, HITS; and (2) develop projects aimed at correcting any identified sanitation deficiencies in these homes to include in the SDS. Through this process, IHS has entered information about hundreds of thousands of eligible Indian homes in HITS and developed thousands of projects in the SDS. According to agency documents, HITS is to include information about each Indian home that is eligible to be served by the Sanitation Facilities Construction program and that may have an existing sanitation deficiency. However, based on our review of IHS documentation and interviews with IHS officials, HITS does not contain all eligible Indian homes that may have existing sanitation deficiencies, and some data in the system are not accurate.

HITS Does Not Contain All Eligible Indian Homes That May Have Existing Water Infrastructure Needs

According to IHS officials, as of February 2018, HITS contained information for about 406,000 Indian homes. However, according to IHS area officials, the system does not contain information about all Indian homes eligible to be served by the Sanitation Facilities Construction program.[59] For example, Oklahoma City Area officials we interviewed estimated that, based on Census data and their professional experience, more than 100,000 Indian homes in their area may be eligible for IHS program assistance but are not included in the system,[60] and an unknown number of those homes likely have existing drinking water and wastewater infrastructure needs.[61] These officials, as well as tribal officials administering the Sanitation Facilities Construction program for their tribes in Oklahoma, said that the system does not contain all eligible Indian homes, in part because it is difficult to identify where tribal members are living since most of the communities in the state are a mixture of tribal and non-tribal residents and are not located on reservations.[62] In addition, Portland Area officials stated that they believe the

---

Tribal Areas: A Report from the Assessment of American Indian, Alaska Native, and Native Hawaiian Housing Needs (Washington, D.C.: January 2017).

[59] According to IHS policy, homes eligible to receive Sanitation Facilities Construction program assistance are 24-hour, year-round family dwellings and are located in one of IHS's service delivery areas. These areas typically include counties encompassing reservations and counties that have a common boundary with a reservation.

[60] As of November 2017, HITS included approximately 155,000 homes within the IHS Oklahoma City Area, which provides assistance to tribes in the states of Oklahoma, Nebraska, and portions of Texas.

[61] Specifically, an IHS Oklahoma City Area official said approximately 70,000 Indian homes are not included in HITS because IHS could not obtain accurate geolocations, but Census data indicate that another 67,000 homes could be eligible to be included in the system.

[62] According to IHS policy, Indian homes that are located in majority non-Indian communities with a population of less than 10,000 are eligible for Sanitation Facilities Construction program assistance. To provide service to

system is missing an unknown number of eligible Indian homes in their area because it is challenging to identify eligible homes that are in scattered locations away from tribal community facilities.[63] In contrast, Navajo Area officials said they believe the system is more than 95 percent complete for their area, in part because the area aligns with the Navajo Nation's lands.

IHS headquarters officials stated that they do not expect HITS to capture all eligible Indian homes, in large part because some tribes have chosen to not provide such information to IHS for cultural or other reasons. These officials said they are focused on working collaboratively with tribes to identify homes that have existing deficiencies rather than all homes eligible for services but added that IHS areas are expected to identify such homes during the normal course of their work. IHS area officials and tribal officials we interviewed stated that identifying eligible Indian homes not located on reservations is resource intensive, and they do not have sufficient resources to locate these homes. IHS Oklahoma City Area officials said it would be helpful to find efficient ways to identify additional eligible homes that may have sanitation deficiencies. For example, these officials said they have started using EPA data to target communities with water systems that do not meet EPA's water quality standards and identify eligible homes within those communities, but they have made limited progress with their existing resources.

Deficiency Levels Are Not Accurate for Every Home in HITS

Standards for Internal Control in the Federal Government calls for management to use quality information to achieve the entity's objectives; such information is appropriate, current, complete, accurate, accessible, and provided on a timely basis.[64] We recognize that it would be resource intensive for IHS to locate every eligible Indian home to include in HITS, but because the system may not contain roughly 20 percent of eligible Indian homes, opportunities exist for IHS to identify in a targeted, efficient way additional homes with existing deficiencies to include in HITS. By implementing a targeted, resource-efficient method to identify additional eligible Indian homes that may have existing sanitation deficiencies to include in HITS, IHS could have better assurance that it has more complete information to help improve its estimate of the number of eligible Indian homes that may need sanitation facilities assistance.

Deficiency levels for thousands of homes may not be accurately captured in HITS. IHS headquarters officials stated that, as of February 2018, of the roughly 406,000 total tribal homes in HITS, about 229,400 homes had a deficiency level of 0. Therefore, the remaining approximately 176,600 tribal homes had deficiency levels 1 through 5. HITS

Indian homes in these communities, IHS contributes the prorated cost to a project that serves the entire community.

[63] IHS Portland Area provides assistance to tribes in the states of Idaho, Oregon, and Washington.

[64] GAO, *Standards for Internal Control in the Federal Government*, GAO-14-704G (Washington, D.C.: September 2014).

automatically assigns a deficiency level 0 to each home when IHS enters it into the system, and homes remain at a deficiency level 0 until IHS develops projects in the SDS to serve those homes. HITS does not provide IHS with the option of recording a home's deficiency level as unassessed, so a deficiency level 0 could indicate that there is no deficiency or that the home has not yet been assessed to determine a deficiency.

IHS area officials we interviewed stated that they were aware of homes with sanitation deficiencies that were not accurately reflected in HITS. For example, Phoenix Area staff said they knew of homes with a deficiency level 4 or 5 that had a deficiency level 0 in HITS because these homes were not yet included within the scope of an SDS project.[65] Also, California Area officials estimated that they had not assessed deficiency levels for about 20,000 eligible homes in their area, and Oklahoma City Area officials said they had not assessed more than 100,000 homes in their area—these homes' deficiency levels all appeared as deficiency level 0 in HITS, but their actual deficiencies were unknown.

According to IHS officials, there are multiple ways to assess homes' deficiency levels. For homes that are not connected to a public water system, such as homes with private wells, IHS staff may need to visit homes to identify any existing deficiencies, with permission of the tribe. For homes connected to a public water system, staff can assign the homes the deficiency level associated with the water system but may need to visit the community to assess the system's overall deficiency level. IHS officials from the California and Oklahoma City areas said they did not have the staff resources to begin the process of identifying whether the deficiency level 0 homes in their areas had deficiencies and developing projects for the SDS to serve them.

IHS headquarters officials stated that they have identified homes that the Sanitation Facilities Construction program has served since implementing HITS in 2015. For example, IHS officials stated that of the about 229,400 homes with a deficiency level 0 in HITS, they had determined that about 90,700 correctly show that deficiency level because they have been included in a project in the SDS since 2015. IHS had not included the remaining approximately 138,700 homes with a deficiency level 0 in a project in the SDS. Therefore, using HITS, IHS could not determine if these homes had (1) no deficiency, (2) a deficiency that IHS addressed prior to 2015, or (3) an unknown deficiency because the homes had not been assessed.

IHS officials stated that in the future they will be able to use HITS to better track the agency's service and project history at the individual home level. However, IHS officials did not explain what steps they would take to identify deficiencies for the approximately 138,700 homes in HITS that had not been included in an SDS project. *Standards for Internal Control in the Federal Government* calls for management to use quality

[65] The IHS Phoenix Area provides assistance to tribes in Arizona, Nevada, and Utah, with the exception of the Navajo Nation, San Juan Southern Paiute Tribe, Tohono O'odham Nation, and the Pascua Yaqui Tribe of Arizona.

information to achieve the entity's objectives; such information is appropriate, current, complete, accurate, accessible, and provided on a timely basis.[66] IHS officials said that improving the system's accuracy would be beneficial. By implementing a mechanism to indicate in HITS whether each home with a deficiency level of 0 has been assessed, IHS could also have more efficient ways to take steps to address the deficiencies of the homes contained in HITS.

Federal Agencies Provided Funding for Tribal Water Infrastructure Projects, but Processes May Not Prioritize Projects That Address the Most Severe Deficiencies

In fiscal year 2016, federal agencies obligated approximately $370 million for tribal drinking water or wastewater infrastructure projects. The agencies with tribal-specific programs for drinking water and wastewater infrastructure—IHS, EPA, and USDA—funded some projects to address what they identified as the most severe sanitation deficiencies— communities and homes that do not have safe drinking water or wastewater disposal facilities. However, the agencies' processes may not always prioritize projects that address the most severe sanitation deficiencies. In addition, during the course of our review, we identified issues with how USDA awarded grants under its Rural Alaska Village Grant program.

Federal Agencies Provided About $370 Million in 2016 for Tribal Water Infrastructure Projects

In fiscal year 2016, federal agencies provided about $370 million to develop, construct, or repair tribal drinking water and wastewater infrastructure projects to address tribes' needs. This amount represents about 11 percent of the more than $3 billion in total existing tribal drinking water and wastewater infrastructure needs that IHS identified in 2016.

Appendix III contains additional detail about agency obligations for tribal drinking water and wastewater infrastructure projects for fiscal years 2012 through 2016.

Federal agency obligations were used to address a variety of tribal drinking water and wastewater infrastructure needs. For example, IHS, EPA, USDA, and the State of Alaska provided approximately $15.9 million for multiple, phased projects to bring first-time, in-home piped drinking water and wastewater service to approximately 90 homes in the

---

[66] GAO-14-704G.

Native Village of Eek in Alaska (see fig. 1). The residents of Eek obtain their drinking water by hauling water from the village washeteria, a building that contains toilets, washing machines, and a spigot for purchasing water for use in the home. Most homes in the community do not have piped water or sewer service to kitchens or bathrooms, and residents use washbasins for handwashing and food preparation and honeybuckets for wastewater disposal. As of April 2017, construction was ongoing, and officials estimated that the entire community of about 300 people would be served by the fall of 2018. See appendix IV for other examples of tribal drinking water and wastewater infrastructure projects that we visited.

Ongoing construction of community drinking water and sewer transmission lines
Source: GAO. | GAO-18-309.

Figure 1. Water Infrastructure Project in Eek, Alaska (April 2017).

In addition to providing financial assistance for projects to design or construct water infrastructure, federal programs provided grants for technical assistance and training for tribal utilities and staff. For example, in fiscal year 2016, USDA awarded a $130,000 grant from its Technical Assistance and Training program to one organization that works with tribes.[67] USDA also awarded a contract to the National Rural Water Association for it to employ a network of technical consultants who can provide on-site technical assistance to eligible systems, including tribally operated systems experiencing day-to-day operational issues, among other challenges.

[67] USDA's Technical Assistance and Training program awards grants to non-profits that provide technical assistance and training to tribes, among other entities. USDA reported that from 2012 through 2016, this program awarded nearly $3 million in such program grants to tribal non-profit entities to help tribes with technical assistance and training needs.

Federal programs mostly did not provide financial assistance for routine operations and maintenance of installed community or individual infrastructure.[68] Tribal officials we interviewed, however, said that paying for operations and maintenance is often the tribe's biggest challenge once a system is constructed or upgraded. For example, officials from one tribe said that the tribe did not have sufficient resources to operate and maintain a newly constructed water treatment system. Tribal officials we interviewed stated that their members are often unable to afford the utility fees needed to support the water system. For private systems, officials from two tribes said some of their members have trouble maintaining new drinking water filtration and septic systems because, for example, the systems are technically complex and costly to maintain. Officials from another tribe said homeowners who have difficulty operating and maintaining a system may return to using an unsafe drinking water source they previously used, for example. According to IHS officials, the agency has been collaborating with EPA, USDA, and tribes to improve collection of information about the cause of some systems' premature failure and to analyze best practices for operations and maintenance of tribal water systems.

Agencies Funded Some Projects to Address the Most Severe Sanitation Deficiencies

Agencies with tribal-specific programs for water infrastructure—IHS, EPA, and USDA—selected and funded projects that address the most severe sanitation deficiencies. Three of these agencies' programs—IHS's Sanitation Facilities Construction, EPA's clean water set-aside, and USDA's Native American program—documented in regulation or policy their goal of funding projects to address these needs. Specifically, according to IHS's Sanitation Facilities Construction program policy, the program's goal includes providing funding first and in greater degree to homes and communities with the greatest needs, that is, those that lack safe drinking water or wastewater disposal, or both.[69] EPA's clean water set-aside program policy states the program's goal is to protect public health in Indian country by addressing the lack of access to sanitation facilities (i.e., deficiency levels 4 and 5 for IHS and EPA).[70] Finally, under the applicable requirements and policy, USDA's Native American program's objective is to provide water and waste disposal facilities and services to low-income rural communities whose residents face significant

---

[68] Some of the programs we reviewed are not authorized to fund operations and maintenance activities, such as EPA's clean water set-aside program, USDA's Native American and Rural Alaska Village Grant programs, and HUD's Indian Community Development Block Grant program.

[69] Indian Health Service, Office of Environmental Health and Engineering, Division of Sanitation Facilities Construction, *Criteria for the Sanitation Facilities Construction Program* (Rockville, MD: 2003).

[70] Environmental Protection Agency, *Clean Water Indian Set-Aside Program Guidance*, EPA-832-B-15-001 (Washington, D.C.: October 2015).

health risks.[71] The program's goal includes funding the neediest projects, giving priority to areas that lack running water, flushing toilets, and modern sewage disposal systems.

Table 2. Indian Health Service's (IHS) scoring factors for ranking projects in the Sanitation Deficiency System (SDS)

Factor	Description	Minimum and maximum points awarded
Health impact	Potential for occurrence of a disease or other adverse human health effect directly attributable to the failure of (or lack of) water or sewer facilities.	0 to 30 points
Project deficiency level	Reflects the deficiency level of facilities to be replaced or modified by the proposed project. Projects with higher deficiency levels receive more points.	0 to 18 points
Capital cost	Relative cost per home served by the project compared to similar projects in the area. Projects with lower cost per home served receive more points.	-20 to 16 points
Operations and maintenance capability	Probability of adequate operations and maintenance of facilities provided through the project.	0 to 16 points
Local tribal priority	Tribe's documented priorities for its preferred projects.	0 to 16 points
Contributions	For projects that leverage funding contributions from non-IHS sources.	0 to 8 points
Factor	Description	Minimum and maximum points awarded
Adequate previous service	For projects that serve communities that have not been provided adequate water and sewage facilities.	0 to 4 points
Local conditions factor	Area can adjust project's overall score to compensate for unusual circumstances, such as project sequencing needs and status of project planning.	-15 to 0 points
Total possible points		108 points

Source: GAO analysis of IHS information. | GAO-18-309.

Notes: According to IHS's guide for reporting sanitation deficiencies, IHS area offices have the discretion to use or not use the following factors when scoring projects: operations and maintenance capability, contributions, and local conditions. Also, IHS area offices may modify the method for scoring local tribal priority points to meet the specific needs of the area. IHS developed its scoring factors for ranking projects in the SDS in 1989 and updated them in 2003.

According to agency policy, IHS's Sanitation Facilities Construction program and EPA's clean water set-aside program prioritize and select projects to fund according to the projects' rankings in each IHS area's SDS list.[72] To create the ranked lists, IHS staff

---

[71] The statute authorizing the section 306C grants makes them available only to communities "whose residents face significant health risks … due to the fact that a significant proportion of the community's residents do not have access to, or are not served by, adequate affordable water supply systems or waste disposal facilities." 7 U.S.C. § 1926c(a)(1). The regulations governing section 306C grants and loans apply to the Native American program. Those regulations identify the objective of the section 306C grants and loans program as providing water and waste disposal facilities and services to low-income rural communities whose residents face significant health risks. 7 C.F.R. § 1777.3. See also U.S. Department of Agriculture, Rural Utilities Service, *Staff Instruction 1780-5: Water and Environmental Program Fund Allocations – Fiscal Year 2017* (Washington, D.C.: September 2017).

[72] EPA's drinking water set-aside program guidance states that EPA regions are to choose how they will identify and select drinking water projects to fund from the drinking water set- aside program, in consultation with tribes in each region. For this program, the regions have selected a variety of approaches. Four EPA regions select projects to fund from the drinking water set-aside program based on their SDS rankings, and the

assign scores to each project based on a set of eight scoring factors, each with a different number of points that may be assigned to a project (see Table 2).

USDA prioritizes and selects projects to fund from its Native American program using a different process than IHS and EPA. USDA's process involves tribes, working with USDA state offices, submitting project grant applications to the headquarters office. USDA state offices score project applications before submitting them to the headquarters office. USDA policy directs the program to make funds available according to priority, and the agency accepts and evaluates applications and awards grants throughout the year. USDA officials said the program maintains a wait list for eligible applications received after all available funds have been obligated each year. According to USDA's scoring sheet for the Native American program, the agency evaluates project applications based on a set of five scoring factors, each with a different number of points to award. These scoring factor categories include population, income, joint financing, and discretionary points that can be awarded at state offices and headquarters (see Table 3). USDA officials said that they also take SDS deficiency levels into account when reviewing project applications, but that the statute authorizing the Native American program does not specifically reference IHS's deficiency level definitions.

Using their respective processes to prioritize and select projects for funding, IHS's Sanitation Facilities Construction program, EPA's clean water set-aside program, and USDA's Native American program obligated a total of nearly $110 million in fiscal year 2016 for projects to meet a mixture of needs. For example, for approximately 190 projects from the SDS that IHS, EPA, and others funded in fiscal year 2016, about 40 percent were projects to address deficiency levels 2 and 3, and about 60 percent were projects to address deficiency levels 4 and 5.[73] Further, in fiscal year 2016, USDA reported that its Native American program funded four projects that provided new drinking water and wastewater service to some tribal communities and funded nine projects that replaced, renovated, or expanded existing infrastructure.[74] Based on our review of IHS and USDA documents, deficiency level 2 and 3 projects as well as replacement and renovation projects can address important water quality and other problems, but they generally do not address the most severe deficiencies or the most significant health risks.

---

remaining five regions generally use an application process to select projects. Because of the variety in how regions implement the drinking water set-aside program, we did not review the extent to which EPA addresses the most severe sanitation deficiencies through this program.

[73] In addition to IHS and EPA, other entities, including tribes, may contribute funds to projects funded from the SDS.

[74] For fiscal year 2016, USDA reported that its Native American program provided funding for 17 drinking water and wastewater infrastructure projects. Of these 17 projects, USDA categorized 13 projects as having the funds' purposes as new, replacement, renovation, or expansion. USDA categorized the remaining 4 projects as having other purposes or did not include a purpose.

Table 3. U.S. Department of Agriculture (USDA) Native American program's scoring factors for evaluating projects to fund

Factor	Description	Minimum and maximum points awarded
Population	Proposed projects are to serve areas with a rural population. Projects serving smaller populations receive more points.	10 to 30 points
Income	Proposed projects are to serve areas with low income. Projects serving populations with lower median household incomes receive more points.	10 to 40 points
Joint financing	Proposed projects that have committed joint financing will receive additional points.	5 to 10 points
State office discretionary	State office director may assign points for items such as natural disasters, to improve compatibility or coordination between the agency and other agencies' selection systems, to assist projects that are most cost effective, high unemployment rate, and severity of health risks, among others.	0 to 15 points
Additional discretionary points	USDA's Rural Utilities Service Administrator may assign additional points for items such as geographic distribution of funds and severity of health risks, among others.	0 to 35 points
Total possible points		130 points

Source: GAO analysis of USDA information. | GAO-18-309.

Based on our review of IHS and EPA documents and interviews with these agencies' officials, we found that their process for prioritizing and selecting projects to fund from the SDS can discourage funding some deficiency level 4 and 5 projects, especially those with a relatively high cost per home. According to some IHS area officials we interviewed, applying IHS's scoring factors and the points associated with each factor means that deficiency level 3 projects may score higher than—and therefore receive funding before—deficiency level 4 or 5 projects. For example, a project's cost per home is a significant contributor to its score because IHS assigns as low as minus 20 points for projects that have a relatively high cost to implement per home. IHS officials said that, typically, deficiency level 3 projects replace existing community infrastructure and serve more homes, which makes those projects' relative cost per home lower than deficiency level 4 and 5 projects. IHS headquarters officials explained that they developed the SDS scoring system in consultation with tribes so the system could balance the need to fund deficiency level 4 and 5 projects with the need to fund projects with lower deficiencies that address health needs and serve a larger number of homes. However, because deficiency level 4 and 5 projects may rank lower than some projects that address less severe deficiencies and rank too low to be funded in a given year, hundreds of feasible projects to address the most severe sanitation deficiencies have remained on SDS lists for 5 years or more, based on our review of these lists. As of the end of fiscal year 2016, many of these projects had not been selected for funding from IHS or EPA.[75]

[75] We reviewed areas' SDS lists from fiscal years 2005 through 2016 and identified approximately 580 feasible deficiency level 4 and 5 projects that appeared in the SDS for 5 or more years. We identified about 90 of these projects that appeared on areas' SDS lists for all 12 years that we reviewed. Projects that do not appear on the

IHS headquarters officials said that the agency is working to improve the extent to which it funds feasible projects to address the highest sanitation deficiencies. For example, these officials said that they are updating the 2003 Sanitation Facilities Construction program guidelines to incorporate subsequently issued guidance, and this update should also more directly align the guidelines with the program's original focus—to prioritize service to Indian homes and communities that lack access to piped water and sewer systems.[76] However, a senior IHS official said that changing the SDS scoring factors is not part of this effort because the current scoring factors balance a number of interests in addition to projects' deficiency levels. The official said that higher deficiency level projects ranking lower than other projects on the SDS list in a given year does not mean that public health needs are going unaddressed. Yet, our analysis shows that projects to address the highest deficiency levels have remained in the SDS for many years. We recognize that IHS faces trade-offs when selecting tribal infrastructure projects to fund. By reassessing the point distribution across the SDS scoring factors as part of IHS's program guidelines update, in light of trade-offs between funding projects that address the most severe sanitation deficiencies and projects that meet other needs, IHS may have better assurance that its projects address the most severe sanitation deficiencies in Indian communities.

Regarding USDA's Native American program, based on our review of agency documents and interviews with USDA officials, we found that the agency's process for prioritizing and selecting projects may not provide USDA with reasonable assurance that it is selecting projects to fund that address the most severe sanitation deficiencies. Specifically, USDA's scoring factors for its Native American program do not include a scoring factor category to evaluate the extent to which projects will address health risks that stem from tribes' lack of drinking water and wastewater infrastructure. In contrast, USDA prioritizes projects to fund under its Colonias grant program using an additional scoring factor that awards points based on the extent to which a proposed project will address health risks stemming from lack of safe drinking water and wastewater disposal in a colonia. For example, USDA awards 50 points for projects in colonias where a lack of access to safe drinking water and wastewater disposal results in a significant health risk. We recommended in December 2009 that USDA take steps to better target its limited funds for the Colonias program, and USDA responded in part by creating the additional scoring factor for colonias to ensure that the neediest colonias receive funding.[77]

---

most recent SDS lists may have received funding at some point during the time frame and were subsequently removed from areas' SDS lists, or they may have remained unfunded as of fiscal year 2016.

[76] In November 2017, IHS notified tribal leaders that the agency planned to consult with tribes on the update to the Sanitation Facilities Construction program guidelines.

[77] GAO-10-126. In July 2012, USDA modified the agency's regulations to allow for additional priority points to projects serving colonias that lack access to water or waste disposal systems and that face significant health problems. 7 C.F.R. § 1777.13(d)(5).

To prioritize Native American program applications that address significant health risks, USDA officials said they use discretionary points. However, according to program policy, USDA state office and headquarters officials may award discretionary points to meet other purposes that are not related to addressing health risks, such as encouraging projects with green infrastructure or promoting geographic diversity among grantees, or they may not award these points at all. As a result, USDA may not have reasonable assurance that it is consistently evaluating or funding project applications in a way that aligns with the Native American program's goal. USDA policy states that both the Native American and Colonias programs are to prioritize areas that lack running water, flushing toilets, and modern sewage disposal systems. By implementing a scoring factor similar to the one in the Colonias program—that is, one that awards points for proposed projects that address health risks from a lack of access to safe drinking water and wastewater disposal—for the Native American program, USDA would have more assurance that it is evaluating project applications consistently and funding projects to address the most severe sanitation deficiencies in Indian communities, consistent with the program's goal.

USDA Did Not Always Award Rural Alaska Village Grants to Authorized Recipients, and the Program's Regulations Are Inconsistent with Its Authority

During the course of reviewing funding for tribal drinking water and wastewater infrastructure projects, we encountered several issues with one of USDA's tribal drinking water and wastewater infrastructure programs, the Rural Alaska Village Grant program.[78] Specifically, section 306D of the Consolidated Farm and Rural Development Act authorizes USDA to make grants to the State of Alaska for the benefit of rural or Native villages to provide for the development and construction of drinking water and wastewater systems.[79] According to USDA reports, these grants are used for projects that have provided, for example, rural Alaska Native residents with access to safe drinking water and flush toilets in their homes. From the program's beginning in fiscal year 1997 through fiscal year 2016, USDA awarded 455 grants totaling more than $444 million to provide safe drinking water and wastewater disposal to thousands of Alaska Natives in remote communities.[80]

---

[78] We did not review whether the grants were used for their intended purpose—the development and construction of water and wastewater systems to improve the health and sanitation conditions in Alaska rural and Native villages. In the course of our review, we did not see any evidence of grants being used for other than their intended purpose.

[79] Section 306D was added to the Consolidated Farm and Rural Development Act by Pub. L. No. 104-127, tit. VII, subtit. B, § 757(a), 110 Stat. 1131 (Apr. 4, 1996) (*codified as amended at* 7 U.S.C. § 1926d(a)).

[80] The Federal Agricultural Improvement and Reform Act of 1996, which amended the Consolidated Farm and Rural Development Act in April 1996, authorizes the program. USDA began awarding Rural Alaska Village grants in fiscal year 1997.

We found that from fiscal year 1997 through fiscal year 2016, USDA awarded 159 Rural Alaska Village grants totaling about $157 million to recipients not authorized by section 306D. These recipients were Alaska Native villages, municipalities, and the Alaska Native Tribal Health Consortium, which is the tribal organization that administers IHS's Sanitation Facilities Construction program in Alaska.[81] USDA's appropriations acts for fiscal years 2012 through 2017, however, authorized USDA to provide Rural Alaska Village grants to the Consortium.[82] Of the 159 grants, USDA awarded 127 grants (about $121 million) to municipalities and Alaska Native villages from 1997 through 2016, and it awarded 32 grants (about $36 million) to the Consortium in 2011 before first receiving authority to do so in fiscal year 2012. Based on our review of a list of USDA grant agreements, selected agreements, and according to agency officials, in 2011, USDA signed 32 such agreements with the Consortium and the communities on whose behalf the Consortium administered the grants. USDA officials stated that the agency made seven total obligations to the Consortium in 2011 for these grants.[83]

USDA officials stated that they did not agree that the agency had awarded Rural Alaska Village grants to ineligible entities because the program's authorizing statute gives the State of Alaska control over the use of the grants, and the state concurred with USDA making some grants directly to other parties. For example, the USDA officials stated that a 2011 memorandum of agreement between USDA, the State of Alaska, IHS, and the Consortium was a vehicle for the state to direct a portion of the Rural Alaska Village grants to other parties. These officials stated that since the statute does not prevent the state from redirecting portions of the grant to other parties, it is not improper for USDA to enter into an agreement with the state to award the grants directly to other parties so that the state does not have to redirect them. In commenting on a draft of this report, USDA noted that the agency has awarded two grants to municipalities since signing the 2011 agreement.

In addition, USDA officials said that they entered into the memorandum of agreement and began awarding grants to the Consortium in 2011 to address problems with the program's administration, which resulted in projects that were delayed or halted. For example, USDA stated in a 2010 report that the State of Alaska had not adequately

---

[81] USDA's implementation of the Rural Alaska Village Grant program has changed over time. From 1997 through 1999, USDA awarded all grants to municipalities and Alaska Native villages; from 2000 through 2005, the agency awarded all grants to the State of Alaska; from 2007 through 2010, USDA awarded grants to municipalities, Alaska Native villages, and the state; and from 2011 through 2016, the agency awarded grants to the Consortium, the state, and two municipalities.

[82] Pub. L. No. 112-55, div. A, tit. III, 125 Stat. 552, 572 (Nov. 18, 2011); Pub. L. No. 113-6, div. A, tit. III, 127 Stat. 198, 217 (Mar. 26, 2013); Pub. L. No. 113-76, div. A, tit. III, 128 Stat. 5, 24 (Jan. 17, 2014); Pub. L. No. 113-235, div. A, tit. III, 128 Stat. 2130, 2153 (Dec. 16, 2014); Pub. L. No. 114-113, div. A, tit. III, 129 Stat. 2242, 2264 (Dec. 18, 2015); Pub. L. No. 115-31, div. A, tit. III, 131 Stat. 135, 157 (May 5, 2017). The Alaska Native Tribal Health Consortium is the name of the consortium formed pursuant to section 325 of Pub. L. No. 105-83, 111 Stat. 1543, 1597-1598 (Nov. 14, 1997).

[83] USDA officials stated that per agency regulations and a memorandum of agreement, the Consortium may bundle projects for multiple eligible Alaska villages and submit them under one request for grant assistance. If approving the request, USDA makes one obligation to the Consortium that is used to support multiple projects.

documented project costs and that USDA staff were concerned that the state had not applied the obligations it received from USDA to the intended communities.[84] According to USDA officials, they have seen a significant improvement in the state's grant administration and more timely delivery of projects since the 2011 agreement. In addition, the Rural Alaska Village Grant program manager said the agency awards grants directly to Native villages that have the capacity to administer them. In commenting on a draft of this report, USDA stated that the agency has made all grants to the Consortium pursuant to the 2011 memorandum of agreement.

The State of Alaska can choose to make subgrants once it receives the Rural Alaska Village grant, but section 306D of the Consolidated Farm and Rural Development Act only authorizes USDA to award grants to the State of Alaska. Moreover, the 2011 memorandum of agreement cannot authorize USDA to award grants to recipients that are not authorized by statute. By ensuring that all Rural Alaska Village grants are awarded only to recipients identified as eligible in section 306D or USDA appropriations acts, USDA will have assurance that it is complying with the law. If USDA wants to award Rural Alaska Village grants to municipalities and Alaska Native villages, it should seek authority to do so, as it did to award such grants to the Consortium.

In addition, the regulations governing the Rural Alaska Village Grant program identify rural or native villages in Alaska as eligible grant recipients.[85] USDA officials explained that the agency amended the Rural Alaska Village Grant program regulations in 2015 to codify the 2011 memorandum of agreement. However, this regulation identifying rural and Alaska Native villages as eligible grant recipients expands USDA's authority to award grants beyond the existing statutory authorities, which do not list rural or Alaska Native villages as eligible recipients.[86] Until USDA amends the Rural Alaska Village Grant program regulations to be consistent with USDA's authority, the agency's regulations will continue to recognize recipients not authorized by statute.

---

[84] U.S. Department of Agriculture, *USDA Report on Streamlining of the Rural Alaska Village Grant Program: Report to Congress* (Washington, D.C.: August 2010).

[85] The initial regulations governing the Rural Alaska Village Grant program identified rural or Native Alaskan [sic] villages as eligible applicants. 62 Fed. Reg. 33462, 33492 (June 19, 1997) (*codified at* 7 C.F.R. § 1780.49(c)(1) (1998)). The current regulations— issued in late 2015— identify a rural or native village in Alaska, Alaska's Department of Environmental Conservation, or the Alaska Native Tribal Health Consortium as eligible recipients. 7 C.F.R. § 1784.8(a).

[86] Agencies cannot issue regulations that confer on the agency any greater authority than that conferred by the governing statute. *See e.g.*, *Ernst & Ernst v. Hochfelder*, 425 U.S. 185, 213-214 (1976) ("the rulemaking power granted to an administrative agency charged with the administration of a federal statute is not a power to make law… [the rule's] scope cannot exceed the power granted the [agency] by Congress"); *Killip v. OPM*, 991 F.2d 1564, 1569 (Fed. Cir. 1993) ("Though an agency may promulgate rules or regulations pursuant to authority granted by Congress, no such rule or regulation can confer on the agency any greater authority than that conferred under the governing statute").

THE EXTENT TO WHICH FEDERAL AGENCIES COLLABORATED TO MEET TRIBES' WATER INFRASTRUCTURE NEEDS VARIED AT THE NATIONAL LEVEL AND IN SIX SELECTED STATES

Most of the seven federal agencies that administer programs to provide drinking water and wastewater infrastructure assistance to Indian tribes have taken actions to collaborate at the national level, and the agencies have identified additional opportunities to collaborate. At the regional level, seven federal agencies we surveyed reported collaborating on a range of activities within six selected states—with some agencies frequently working together and others rarely collaborating—and the agencies identified opportunities to increase collaboration at the regional level to better serve tribes.

Most Reviewed Agencies Have Taken Actions to Collaborate at the National Level on Tribal Water Infrastructure and Have Identified Additional Opportunities to Increase Collaboration

Most of the seven federal agencies we reviewed have taken actions to collaborate at the national level and identified additional opportunities to collaborate that they have not yet taken. In our previous work, we found that achieving important national outcomes—such as providing access to safe drinking water and wastewater disposal—often requires coordinated and collaborative efforts of a number of programs spread across the federal government.[87] For example, IHS, EPA, USDA, HUD, and Reclamation have formed a national tribal infrastructure task force to facilitate the agencies' collaborative efforts when providing services, support, and technical assistance to tribes.[88]

The tribal infrastructure task force's efforts reflect some of the key features that we have found all collaborative mechanisms benefit from in our previous work:[89]

- Written guidance and agreements. We have previously reported that agencies that articulate their agreements in documents can strengthen their commitment to working collaboratively.[90] The members of the tribal infrastructure task force first documented their agreement in a memorandum of understanding in 2007, the year the task force was created. The agencies updated the memorandum most recently in 2013, and they use the document to formally agree on the group's common goal and purposes and to clarify and define roles and responsibilities.

---

[87] GAO, *Fragmentation, Overlap, and Duplication: An Evaluation and Management Guide*, GAO-15-49SP (Washington, D.C.: Apr. 14, 2015), and GAO-17-559.

[88] The Department of Health and Human Services represents IHS on the task force, and Interior represents its bureaus, including Reclamation.

[89] GAO-12-1022.

[90] GAO-12-1022.

Having participating agencies document their agreements on how they will be collaborating, and continually updating and monitoring these agreements, are practices that are consistent with our prior work.[91]

- Outcomes and accountability. In our previous work, we have reported on the importance of groups having clear goals.[92] In its 2013 memorandum of understanding, the tribal infrastructure task force identified a common goal of improving access to safe drinking water and basic sanitation for American Indians and Alaska Natives. In the memorandum, the member agencies also agreed on the task force's stated purposes, one of which is to enhance the efficient leveraging of funds.[93]
- Leadership. We have found that identifying one agency as the leader of a collaborative group is often beneficial because it centralizes accountability and can speed decision making.[94] EPA has served as the federal focal point for the task force; this has included hosting the task force's website that serves as a common source for documents the group has produced. According to an official involved with the task force, EPA's role has provided continuity in leadership.

The task force's efforts have yielded some specific results. For example, in 2013, tribal infrastructure task force members agreed to adopt a uniform preliminary engineering report template, a key supporting document that multiple agencies require in their project application and evaluation processes. Task force members created this template in part in response to our October 2012 recommendation that EPA and USDA develop such a document.[95] According to agency officials we interviewed, the report template has been helpful for tribes since they no longer have to produce different versions of the same document when submitting multiple applications to different agencies. USDA officials said they have since worked with other agencies to develop an online version of the preliminary engineering report that is accessible to task force members and others to further improve collaboration.

---

[91] GAO-12-1022.

[92] GAO-12-1022.

[93] According to the 2013 memorandum of understanding, the task force's other purposes are to: facilitate a common understanding of relevant programs and policies of each agency; improve American Indian and Alaska Native communities' capacity to operate and maintain sustainable infrastructure; work collaboratively with tribes to promote an understanding of available federal programs; identify ways to improve planning, construction, operating and maintaining infrastructure; and continue the necessary procedures and structures to facilitate the exchange of information.

[94] GAO-12-1022.

[95] We recommended that EPA and USDA ensure the timely completion of an interagency effort to develop guidelines to assist states in developing their own uniform preliminary engineering reports to meet federal and state requirements. GAO, *Rural Water Infrastructure: Additional Coordination Can Help Avoid Potentially Duplicative Application Requirements*, GAO-13-111 (Washington, D.C.: Oct. 16, 2012). EPA and USDA, along with IHS, HUD, and Interior, developed and adopted the uniform preliminary engineering report template as members of the tribal infrastructure task force.

However, according to agency officials involved with the task force, there may be additional opportunities to improve the efficiency of their collaboration at the national level. For example, in 2011, a workgroup of the task force identified a series of 10 options to increase the efficiency of collaboration by streamlining their application processes.[96] As of November 2017, according to agency officials, the task force had not acted on most of the options. One such option was for agencies to better align their different funding and application cycles where possible.

Several tribal officials and representatives from a tribal organization we interviewed cited challenges with complying with the agencies' different application requirements. For example, they said that doing so can be resource intensive and can make it difficult to obtain funds for one project. Other tribal officials we interviewed also identified ways that agencies could improve their collaboration that would benefit tribal applicants and that the task force did not identify in its 2011 report. For example, various tribal officials suggested that agencies standardize federal program application processes and coordinate their outreach to tribes to discuss agency programs and their requirements.

According to an agency official involved with the task force, when the group considered which options from the 2011 report to implement, member agencies focused their efforts on implementing those that were most achievable given the agencies' limited resources. Other officials also said that it would be worthwhile to reconsider some of the options identified in the report. As stated in the task force's 2013 memorandum of understanding, one of its purposes is to enhance the member agencies' efficient leveraging of funds. By reviewing the 2011 task force report and identifying and implementing additional actions to help increase their collaboration, the task force member agencies could improve their ability to leverage limited program funds.

Federal Agencies' Regional Offices Collaborated to Varying Extents within Six Selected States and Reported That Additional Collaboration Would Be Beneficial

The regional offices of the seven federal agencies we surveyed collaborated with each other to varying extents in the six selected states (Alaska, Arizona, California, New York, Oklahoma, and South Dakota).[97] In the 2013 memorandum of understanding, the tribal infrastructure task force member agencies—IHS, EPA, USDA, HUD, and

[96] Federal Infrastructure Task Force on Tribal Access to Safe Drinking Water and Basic Sanitation, Streamlining Preconstruction Paperwork Workgroup, *Overview of Tribal Infrastructure Funding Application Processes and Recommended Streamlining Opportunities* (Washington, D.C.: February 2011).

[97] The seven federal agencies organize their field structures differently, with some using region, district, area, or state offices as the entity that works with tribes on water infrastructure projects. We refer to all of these office types as regional offices in this discussion. We also sent a survey to the State of Alaska because the state provides a 25 percent match for two federal water infrastructure programs. We did not include other state agencies in our survey because they do not provide matching funds for federal tribal programs.

Reclamation— agreed that they are expected to collaborate at the regional level to achieve a common goal of providing safe drinking water and basic sanitation for tribes. However, based on our review of agency survey responses, these agencies did not always collaborate in each of the six states. We measured agency collaboration in terms of the number of instances in which one agency regional office reported using a collaborative mechanism with another agency.[98] These collaborative mechanisms include state- or project-level working groups, memorandums of understanding, and shared databases, among others. In responses to our survey, we found that the number of instances of agency regional offices reporting that they used one or more collaborative mechanism with other agencies varied across the six states. For example, the agencies' regional offices collaborated the most in Alaska and the least in New York and Oklahoma.[99] Figure 2 shows the percentage of instances where an agency reported using a collaborative mechanism with another agency when jointly working on tribal drinking water and wastewater activities in the six states, out of the total possible instances. Appendix II contains additional details about our survey and agency responses.

In responses to our survey, certain agencies' regional offices reported collaborating with each other in some states but not in other states.[100] For example, EPA and USDA regional offices both reported working together in Alaska, Arizona, and California, but not in New York, Oklahoma, or South Dakota. IHS and HUD reported collaborating with each other in three states but not in the other states. Not all agencies work with tribes in every state. For example, Reclamation does not operate in Alaska or New York, so we did not survey the agency in those states. The Corps' regional offices responded that they are not authorized to work on drinking water or wastewater infrastructure with tribes in two of the six states. In contrast, IHS and EPA regional offices reported collaborating with each other in all six states, the most of any agency pair.

[98] In our survey, for each state, we asked each agency's regional office whether it had collaborated with the other agencies in that state during the 3 years prior to the survey on tribal drinking water or wastewater activities using the following mechanisms: (1) memorandum of understanding or agreement, (2) interagency agreement to transfer funding, (3) working group/task force/committee, (4) consulting on project selection, (5) sharing project documents, (6) geographic co-location, (7) shared database or other data sharing, (8) conferences/forums, (9) informal or ad hoc communications, and (10) personnel detailing or sharing. We also asked the agencies' regional offices about the specific drinking water and wastewater activities that they worked on with the other agencies in each state. Appendix II contains information about the responses to those questions.

[99] We previously found that federal agencies frequently use more than one mechanism to address a complex issue. Mechanisms differ in complexity and scope but we found that they all benefit from certain key features, including, for example, having identified resources needed to initiate or sustain the collaborative effort. See GAO-12-1022.

[100] We are considering that a pair of agencies' regional offices collaborated if both agencies' regional offices reported that they had used at least one mechanism to collaborate with each other during the 3-year period covered by the survey.

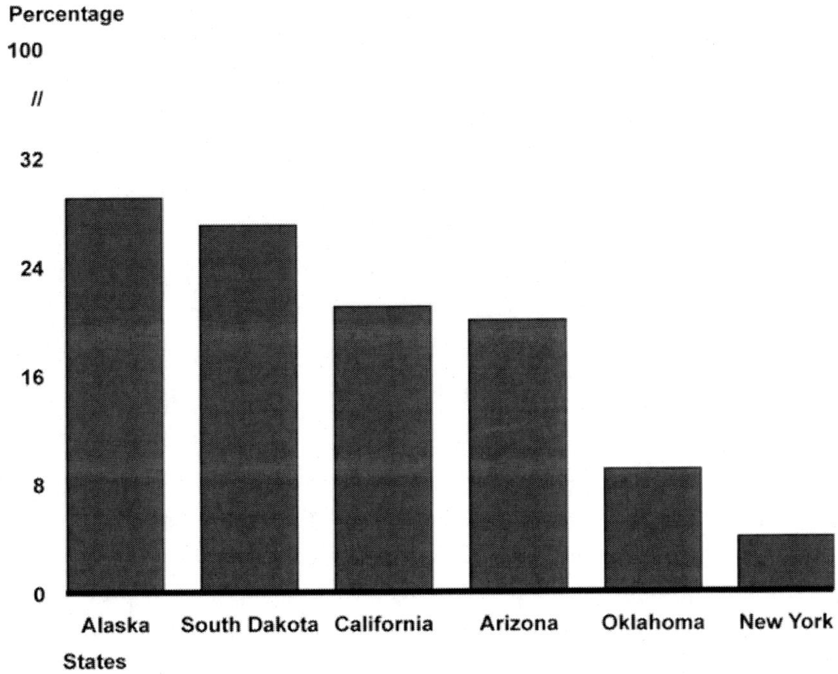

Source: GAO analysis of survey responses. | GAO-18-309.

Notes: We surveyed regional offices of seven federal agencies: Bureau of Reclamation, Department of Housing and Urban Development, Economic Development Administration, Environmental Protection Agency, Indian Health Service, U.S. Army Corps of Engineers, and U.S. Department of Agriculture. We also included the Alaska Department of Environmental Conservation in our survey in Alaska. We surveyed all seven federal agencies in Arizona, California, Oklahoma, and South Dakota (Continued …)

In New York, we surveyed six federal agencies because Reclamation does not operate in New York. In Alaska, we surveyed six federal agencies and the Alaska Department of Environmental Conservation because Reclamation does not operate in Alaska

In our survey, for each state, we asked each agency's regional office whether it had collaborated with the other agencies in that state on tribal drinking water or wastewater activities using the following mechanisms: (1) memorandum of understanding or agreement, (2) interagency agreement to transfer funding, (3) working group/task force/committee, (4) consulting on project selection, (5) sharing project documents, (6) geographic co-location, (7) shared database or other data sharing, (8) conferences/forums, (9) informal or ad hoc communications, and (10) personnel detailing or sharing. To compare the agencies' responses across the six states, we calculated the number of instances that an agency reported using a mechanism with another agency in each state, out of the maximum possible yes responses in each state. The time frame covered by this figure is approximately May 2014 through April 2017. We disseminated the survey in May 2017 and asked agencies to report about their collaboration during the 3-year period prior to our survey.

Figure 2. Agencies' Reported Use of Collaborative Mechanisms to Jointly Work on Tribal Drinking Water and Wastewater Infrastructure Activities in Six States.

In their responses to our survey and in interviews, the seven federal agencies' regional offices most frequently identified three key factors that limited how much they collaborated in the six states.[101] Specifically:

- Incompatibility of agency policies and missions. Agencies' regional offices reported that having incompatible policies or agency missions was a factor that had hindered their collaboration with other agencies. For example, IHS and HUD regional offices in four states reported that a restriction on IHS's ability to serve new homes constructed with grants from HUD's housing programs limited their collaboration.[102] Several agencies' regional offices reported that having compatible policies helped their collaboration. For example, IHS and USDA regional offices in Alaska responded that multiple agencies' use of IHS's SDS list as a common source for identifying potential projects to fund has helped collaboration. We previously found that adopting compatible policies and procedures is one way for agencies to establish means of operating across agency boundaries.[103]
- Insufficient resources. An additional factor that hindered agencies' collaboration was insufficient staff and financial resources. For example, HUD and IHS regional officials we interviewed in Arizona said that a state-level tribal infrastructure working group they were involved in became inactive in 2016, after the lead agency determined it was unable to continue dedicating staff resources to that role and none of the other agencies picked up the lead. In contrast, several IHS and EPA regional offices reported that the existence of standard interagency agreements that facilitate transferring EPA funds to IHS helped them collaborate and leverage funding for projects that each agency would not have funded on its own. Identifying and leveraging the resources needed to initiate or sustain a collaborative effort is a key consideration for implementing the interagency collaborative mechanisms we previously identified.[104]
- Absence of personal relationships. Agencies' regional offices also reported that the absence of relationships with staff from other agencies hindered their collaboration. In contrast, agencies' regional offices reported that having good

---

[101] In each state, the agencies' regional offices reported on the factors that hindered and helped their collaboration with every other agency's regional office we surveyed in that state. We then counted the frequency with which the agencies reported different factors. We use the term several to indicate where more than two agencies' regional offices reported that a factor hindered or helped their collaboration.

[102] New homes constructed with grants from HUD's housing programs are not eligible to be served by the Sanitation Facilities Construction program because IHS's annual appropriations acts have prohibited the use of IHS appropriations for sanitation facilities construction for new homes funded with HUD housing grants. *See e.g.*, Pub. L. No. 112- 74, div. E, tit. III, 125 Stat.786, 1029 (Dec. 23, 2011); Pub. L. No. 115-31, div. G, tit. III, 131 Stat. 135, 485-486 (May 5, 2017).

[103] GAO-12-1022.

[104] GAO-12-1022.

working relationships with staff from other agencies helped their collaboration. For example, USDA's regional office and the State of Alaska reported that their strong relationships with each other and other agencies in Alaska helped their collaboration, and that these relationships were enhanced by agency staff's frequent communication through regular meetings. We previously found that having positive working relationships can bridge organizational cultures and build trust.[105]

In their responses to the survey and in interviews, several agencies' regional offices identified examples of inefficiencies that have occurred when they did not collaborate, including inefficient use of their limited resources. For example, officials from one EPA region we interviewed said that there have been years when EPA staff spent time developing a project only to learn that USDA had already funded the same project. The officials stated that this inefficiency could have been avoided if they had been communicating with their USDA counterparts about the projects that each agency was considering to fund. Also, in two states, EPA's regional office reported that EPA and USDA may be missing opportunities to leverage funding for individual projects by not sharing information about projects.

In all six states, nearly all of the federal agency regional offices responded that it would be beneficial to increase their collaboration. Specifically, more than 90 percent of the federal agency respondents identified at least one collaborative mechanism that would be beneficial for them to begin using with another agency.[106] The specific mechanisms that the agencies identified appeared to relate to the amount of collaboration in which they had already engaged. For example, agency regional offices that reported not having collaborated with another agency most frequently said that it would be beneficial to begin having informal communications with their counterparts in other agencies and to start sharing project-specific documents such as preliminary engineering reports. Alternatively, agency regional offices that reported having collaborated with another agency most frequently responded that it would be beneficial to begin using a memorandum of understanding as an additional mechanism for collaborating where they had not already done so. In the tribal infrastructure task force's 2013 memorandum of understanding, the member agencies—IHS, EPA, USDA, HUD, and Reclamation—agreed that they are expected to collaborate at the regional level to provide safe drinking water and basic sanitation for tribes and to more efficiently leverage program funds. By directing their regional offices to identify and pursue additional mechanisms to increase their collaboration, the task force member agencies would have better assurance that their

---

[105] GAO-12-1022.

[106] In our survey, for every collaborative mechanism that an agency said it had not used with the other agencies in each state, we asked whether it would be beneficial to use that mechanism in the future. Appendix II has additional information about the mechanisms that agencies reported would be beneficial to use in the future.

regional offices are efficiently leveraging limited program funds and following through on the commitment to collaborate.

CONCLUSION

Identifying and addressing drinking water and wastewater infrastructure needs in Indian country is a difficult undertaking. IHS dedicates a significant effort each year to working with tribes to identify their existing drinking water and wastewater infrastructure needs. However, one of IHS's systems—HITS—may be missing tens of thousands of eligible Indian homes, an unknown number of which may have existing sanitation deficiencies. Additionally, some homes' deficiency levels in HITS are inaccurate. By implementing a targeted, resource-efficient method to identify additional eligible Indian homes that may have existing sanitation deficiencies to include in HITS, IHS could have better assurance that it has more complete information to help improve its estimate of the number of eligible Indian homes that may need sanitation facilities assistance. Also, IHS officials said that improving the system's accuracy would be beneficial. By implementing a mechanism to indicate in HITS whether each home with a deficiency level of 0 has been assessed, IHS could also have more efficient ways to take steps to address the deficiencies of the homes contained in HITS.

IHS and USDA funded some projects to address the most severe sanitation deficiencies, but residents of many Indian homes remain without safe drinking water or wastewater disposal as the agencies also prioritized and funded projects that addressed other needs. We recognize that IHS faces trade-offs when selecting tribal infrastructure projects to fund. By reassessing the point distribution across the SDS scoring factors as part of IHS's program guidelines update, in light of trade-offs between funding projects that address the most severe sanitation deficiencies and projects that meet other needs, IHS may have better assurance that its projects address the most severe sanitation deficiencies in Indian communities. Also, by USDA implementing a scoring factor similar to the one in its Colonias program—that is, one that awards points for proposed projects that address health risks from a lack of access to safe drinking water and wastewater disposal—for the Native American program, USDA would have more assurance that it is evaluating project applications consistently and funding projects to address the most severe sanitation deficiencies in Indian communities, consistent with the program's goal.

USDA has provided thousands of Alaska Natives with safe drinking water and wastewater infrastructure through its Rural Alaska Village Grant program. However, USDA awarded some grants to recipients not authorized by statute. By ensuring that all Rural Alaska Village grants are awarded only to recipients authorized by statute, USDA will have assurance that it is complying with the law. If USDA wants to award Rural

Alaska Village grants to municipalities and Alaska Native villages, it should seek authority to do so as it did to award these grants to the Alaska Native Tribal Health Consortium. Also, until USDA amends the Rural Alaska Village Grant program regulations to be consistent with USDA's authority, the agency's regulations will continue to recognize recipients not authorized by statute.

The five agencies that participate in the national tribal infrastructure task force have committed to working together at the national and regional levels to increase tribes' access to safe drinking water and basic sanitation. In our previous work, we have found that achieving important national outcomes, such as providing access to safe drinking water and wastewater disposal, often requires collaborative efforts by a number of programs across the federal government. At the national level, the task force has not acted on most of the options it previously identified to improve member agencies' collaboration. By reviewing the 2011 task force report and identifying and implementing additional actions to help increase their collaboration, the task force member agencies could improve their ability to leverage limited program funds. At the regional level, we found that the task force member agencies had not fulfilled their commitment to collaborate in all of the six states we reviewed. Responses to our survey also indicated that there is unrealized potential for the task force member agencies' regional offices to increase the extent of their collaboration. By directing their regional offices to identify and pursue additional mechanisms to increase their collaboration, the task force member agencies would have better assurance that their regional offices are leveraging limited program funds and following through on their commitment to collaborate.

RECOMMENDATIONS FOR EXECUTIVE ACTION

We are making 16 recommendations—two to IHS to improve information in HITS; one each to IHS and USDA to review their project selection processes; two to USDA to address issues with its Rural Alaska Village Grant program; and two each to IHS, USDA, EPA, HUD, and Reclamation to increase collaboration at the national and regional levels.

- The Director of IHS should implement a targeted, resource-efficient method to identify additional eligible Indian homes that may have existing deficiencies to include in HITS. (Recommendation 1)
- The Director of IHS should implement a mechanism to indicate in HITS whether each home with a deficiency level of 0 has been assessed. (Recommendation 2)
- The Director of IHS should reassess the point distribution across the SDS scoring factors as part of its program guidelines update, in light of trade-offs between funding projects that address the most severe sanitation deficiencies and projects that meet other needs. (Recommendation 3)

- The Assistant to the Secretary of Agriculture for Rural Development should implement a scoring factor that awards points for proposed Native American program grant projects that address health risks from a lack of access to safe drinking water and wastewater disposal, as it does with the Colonias grant program. (Recommendation 4)
- The Assistant to the Secretary of Agriculture for Rural Development should ensure that all Rural Alaska Village grants are awarded only to recipients authorized by law or seek authority to award grants to municipalities and Alaska Native villages. (Recommendation 5)
- The Assistant to the Secretary of Agriculture for Rural Development should amend the Rural Alaska Village Grant program regulations so that they are consistent with USDA's authority. (Recommendation 6)
- The Director of IHS, in cooperation with other members of the tribal infrastructure task force, should review the 2011 task force report and identify and implement additional actions to help increase the task force's collaboration at the national level. (Recommendation 7)
- The Administrator of EPA, in cooperation with other members of the tribal infrastructure task force, should review the 2011 task force report and identify and implement additional actions to help increase the task force's collaboration at the national level. (Recommendation 8)
- The Assistant to the Secretary of Agriculture for Rural Development, in cooperation with other members of the tribal infrastructure task force, should review the 2011 task force report and identify and implement additional actions to help increase the task force's collaboration at the national level. (Recommendation 9)
- The Deputy Assistant Secretary of the Department of Housing and Urban Development's Office of Native American Programs, in cooperation with other members of the tribal infrastructure task force, should review the 2011 task force report and identify and implement additional actions to help increase the task force's collaboration at the national level. (Recommendation 10)
- The Commissioner of Reclamation, in cooperation with other members of the tribal infrastructure task force, should review the 2011 task force report and identify and implement additional actions to help increase the task force's collaboration at the national level. (Recommendation 11)
- The Director of IHS, in cooperation with other members of the tribal infrastructure task force, should direct IHS area offices to identify and pursue additional mechanisms to increase their collaboration. (Recommendation 12)
- The Assistant to the Secretary of Agriculture for Rural Development, in cooperation with other members of the tribal infrastructure task force, should

direct USDA state offices to identify and pursue additional mechanisms to increase their collaboration. (Recommendation 13)

AGENCY COMMENTS AND OUR EVALUATION

- The Administrator of EPA, in cooperation with other members of the tribal infrastructure task force, should direct EPA regional offices to identify and pursue additional mechanisms to increase their collaboration. (Recommendation 14)
- The Deputy Assistant Secretary of the Department of Housing and Urban Development's Office of Native American Programs, in cooperation with other members of the tribal infrastructure task force, should direct HUD regional offices to identify and pursue additional mechanisms to increase their collaboration. (Recommendation 15)
- The Commissioner of Reclamation, in cooperation with other members of the tribal infrastructure task force, should direct Reclamation regional offices to identify and pursue additional mechanisms to increase their collaboration. (Recommendation 16)

We provided a draft of this report for review and comment to the Department of Health and Human Services (for IHS), HUD, the Department of the Interior (for Reclamation), EPA, USDA, the Department of Defense (for the Corps), and the Department of Commerce (for EDA). Of the five agencies to which we directed recommendations, three— Health and Human Services, HUD, and Interior—agreed with the recommendations directed to them. The fourth agency, EPA, agreed with one of the recommendations and agreed with the intent of the second recommendation but proposed revised language, as discussed below.

The Acting Director of Grants Evaluation for HUD's Office of Native American Programs provided comments by e-mail, and Health and Human Services, Interior, and EPA provided written comments that are reproduced in appendixes V, VI, and VII, respectively. The fifth agency to which we directed recommendations, USDA, disagreed with the two recommendations regarding the Rural Alaska Village Grant program and neither agreed nor disagreed with the other three recommendations directed to it, although the agency proposed alternative language for two of these recommendations in its written comments, reproduced in appendix VIII. Of the two agencies to which we did not direct recommendations, Defense provided a letter, reproduced in appendix IX, in which it indicated the agency had no comments on the report, and Commerce's Audit Liaison stated in an e-mail that Commerce would not send a formal comment letter. In addition, Health and Human Services, USDA, and EDA (for Commerce) provided technical comments, which we incorporated in the report as appropriate.

In its written comments, EPA requested that we revise the language of the recommendation that the members of the tribal infrastructure task force direct their regional, state, or area offices to identify and pursue additional mechanisms to increase their collaboration. EPA stated that it agreed with the intent of the recommendation but that it was concerned that, as worded, the recommendation may not achieve the intended goal. Instead, EPA stated that it can accomplish increased regional collaboration through multiple avenues and as such, provided revised language that would remove reference to its regional offices taking the recommended action. We encourage EPA to take advantage of increasing regional collaboration through all avenues it sees fit. However, because EPA's regional offices are the entities that collaborate with other agencies in the various regions, we continue to believe it is important for these offices to participate in identifying and implementing the means for increasing collaboration in their respective regions. As a result, we did not modify the recommendation language in response to EPA's comment.

In its written comments, USDA stated it disagreed with our statements concerning the Rural Alaska Village Grant program and asked that we remove the two corresponding recommendations from our report.

Specifically, USDA stated that our recommendations are unnecessary because the agency is operating within its authorities. USDA stated that it believes providing grants directly to parties other than the state— including Alaska Native villages and municipalities—under the 2011 memorandum of agreement is consistent with the purpose of section 306D of the Consolidated Farm and Rural Development Act and appropriations made for the program. As we state in the report, we agree that the State of Alaska can choose to make subgrants once it receives the Rural Alaska Village grant. We also state in the report that we did not see any evidence of grants being used for other than their intended purposes during the course of our review. However, the language of section 306D only authorizes USDA to award grants to the State of Alaska and not directly to other entities. Therefore, we believe that our recommendations are necessary. If USDA wants to make Rural Alaska Village grants to municipalities and Alaska Native villages, it should seek authority to do so as it did to award such grants to the Alaska Native Tribal Health Consortium.

Regarding our fourth recommendation that USDA implement a scoring factor that awards points for proposed Native American program projects that address health risks, USDA stated that it would like clarification as to what form of scoring factor would be acceptable to address this recommendation. USDA stated that it would prefer to use its discretionary points under the program's existing regulations to award additional points to give a higher priority to projects that address a lack of access to safe drinking water and wastewater disposal, and that the agency could implement this change at the start of fiscal year 2019 or sooner. In contrast, USDA stated that changing the program's regulations to implement the scoring factor could take 18 months or longer. USDA also stated that this approach would only have a programmatic effect in fiscal years when demand for Native American program grant funds exceeds the available funding. Our intent for the recommendation as written is to provide USDA with the flexibility to best determine how to implement it. If USDA has determined that using its discretionary points under the program's existing regulations gives greater priority to addressing health risks faced by Native American communities, and that such an approach is consistent with applicable law, such an approach could meet the intent of our recommendation.

USDA also requested in its written comments that we modify the language of the ninth recommendation aimed at increasing collaboration at the national level by removing reference to increasing national collaboration and that we modify the thirteenth recommendation aimed at increasing regional collaboration by removing reference to the agency's state offices and regional level collaboration. USDA did not provide a clear rationale for its requested change for either recommendation. We continue to believe that implementing these recommendations, as worded, would help improve collaboration at the national and regional levels. Therefore, we did not modify the language in response to USDA's comments.

In several places in its written comments, USDA stated that our draft either omitted information or contained inaccurate information and requested that we make modifications. Specifically, USDA stated that we omitted statutory language for the Native American program in a few places in the report. In response, we added additional language from and about the Native American program's authorizing statute in several places. USDA also stated the report is missing information about the scope of some of its programs, including its Technical Assistance and Training program. In response, we added more information about this program, including obligations made to non-profit organizations that work on behalf of tribes. Further, USDA stated that we did not accurately characterize certain activities that USDA conducts under some of its programs, including identifying tribal needs and conducting operations and maintenance. In response, we modified language to reflect additional information about how USDA identifies tribal needs and to indicate that the Native American program is not authorized to fund operations and maintenance. Regarding the Rural Alaska Village Grant program, USDA stated that we did not accurately represent information shared by a USDA official and information about the number of grants made to the Consortium. We revised language attributed to the official and clarified information about the number of grants awarded based on additional information that USDA provided by e-mail after submitting its written comments.

In other cases where USDA requested revisions to the draft in its written comments, we did not make suggested changes because they did not align with the scope of our review. Specifically, in addition to its Technical Assistance and Training program, USDA asked that we add information about tribal obligations under its Solid Waste Management program.

Since federal agency efforts to fund solid waste management projects are outside the scope of this review, we did not make this revision. In addition, USDA requested that we limit our discussion of Rural Alaska Village Grant awards to fiscal year 2011 and forward. We did not make this change because USDA's grants to municipalities and Native villages prior to 2011 are directly relevant to our findings and are within the scope of this review. Finally, USDA asked that we edit our description of the findings of a 2010 report to Congress by citing a different report instead. We did not make this change because the original report contained relevant information for our findings.

As agreed with your offices, unless you publicly announce the contents of this report earlier, we plan no further distribution until 30 days from the report date. At that time, we will send copies to the appropriate congressional committees, the Secretary of Agriculture, the Secretary of Commerce, the Secretary of Defense, the Secretary of Health and Human Services, the Secretary of Housing and Urban Development, the Secretary of the Interior, the Administrator of the Environmental Protection Agency, and other interested parties.

Anne-Marie Fennell
Director, Natural Resources and Environment

J. Alfredo Gómez
Director, Natural Resources and Environment

List of Requesters

The Honorable Frank Pallone, Jr.
Ranking Member
Committee on Energy and Commerce
House of Representatives

The Honorable Paul Tonko
Ranking Member Subcommittee on Environment
Committee on Energy and Commerce
House of Representatives

The Honorable Jared Huffman
Ranking Member
Subcommittee on Water,
Power and Oceans Committee on Natural Resources
House of Representatives

The Honorable Grace Napolitano
Ranking Member
Subcommittee on Water Resources and Environment Committee
on Transportation and Infrastructure
House of Representatives

APPENDIX I: OBJECTIVES, SCOPE, AND METHODOLOGY

The objectives of our review were to examine the extent to which the seven federal agencies, as applicable, (1) identified Indian tribes' drinking water and wastewater infrastructure needs; (2) funded tribal drinking water and wastewater infrastructure projects, including projects to address the most severe sanitation deficiencies; and (3) collaborated to meet Indian tribes' drinking water and wastewater infrastructure needs.

To address these objectives, we reviewed our previous reports, other agency reports, and agency obligations to identify the federal agencies that provide financial or other assistance to Indian tribes for drinking water and wastewater infrastructure. We identified seven agencies, as shown in Table 4. We identified the Indian Health Service (IHS), Environmental Protection Agency (EPA), and U.S. Department of Agriculture (USDA) as federal agencies that have drinking water and wastewater infrastructure programs specifically targeted to provide financial assistance for planning and construction to address Indian tribes' needs. According to IHS documentation, such needs arise from a sanitation deficiency in existing drinking water or wastewater infrastructure (or lack thereof) that can negatively affect public health. In addition, the Department of Housing and Urban Development (HUD), the Department of the Interior's Bureau of Reclamation, the Department of Commerce's Economic Development Administration (EDA), and the U.S. Army Corps of Engineers administer programs that may assist tribes with drinking water and wastewater infrastructure planning and construction. The types of assistance these agencies provide vary by program, and each program has its own eligibility requirements and authorities.

To determine the extent to which these federal agencies identified Indian tribes' drinking water and wastewater infrastructure needs, we identified requirements for IHS and EPA to collect and report information on needs, but we did not identify such requirements for the other agencies.[107] We reviewed IHS's project-level tribal drinking water and wastewater infrastructure needs data from the Sanitation Deficiency System (SDS) for fiscal year 2016, the most recent year of data available at the time of our review. The SDS contains information about proposed drinking water and wastewater infrastructure projects, including each project's estimated cost. IHS policy directs area staff to invite all federally recognized tribes to identify potential projects each year. Area staff then work with interested tribes to develop projects and enter project information into standard fields in the SDS. As of the end of fiscal year 2016, the SDS included more than 2,000 projects for 373 tribes. We also reviewed IHS's most recent reports describing tribal drinking water and wastewater needs.

Table 4. Key federal programs we reviewed that provide assistance to tribes for drinking water and wastewater infrastructure

Agency	Program
U.S. Department of Agriculture's Rural Utilities Service	Native American Rural Alaska Village Grant Water and Waste Disposal Loan and Grant
U.S. Department of Commerce's Economic Development Administration	Public Works Economic Adjustment Assistance
U.S. Department of Defense's Army Corps of Engineers	Specific authorizations for projects in laws, including Environmental Infrastructure Program
U.S. Department of Health and Human Services' Indian Health Service	Sanitation Facilities Construction
U.S. Department of Housing and Urban Development	Indian Community Development Block Grant
U.S. Department of the Interior's Bureau of Reclamation	Specific authorizations for projects in laws, including enacted Indian water rights settlements
Environmental Protection Agency	Clean Water Indian Set-Aside Drinking Water Infrastructure Grants Tribal Set-Aside Alaska Native Villages and Rural Community Water Grant

Source: GAO analysis of agency documents. | GAO-18-309.

In addition, we reviewed information about tribal public drinking water systems reported in EPA's 2013 Drinking Water Infrastructure Needs Survey and Assessment report. EPA assesses and reports on the nation's public water systems' capital improvement needs every 4 years, including needs of tribally owned or operated drinking water systems. For its 2013 report, EPA assessed tribal water system needs by administering a survey to a statistical sample of 306 tribal water systems out of 956 identified tribal public drinking water systems.

[107] In conducting this work, we relied on the concept of needs as defined by IHS and EPA in their reports as opposed to independently defining the concept of need or evaluating the legitimacy of the reported needs. According to IHS, needs arise from a sanitation deficiency in existing drinking water or wastewater infrastructure (or lack thereof) that can negatively affect public health. According to EPA, its estimates of needs represent infrastructure projects necessary for water systems to continue to provide safe drinking water to the public.

We assessed the reliability of SDS project-level needs data and information from EPA's 2013 Drinking Water Infrastructure Needs Survey and Assessment report by reviewing our previous related work regarding the use of these data and documentation from IHS and EPA. We also interviewed IHS and EPA officials involved with identifying tribal water needs from headquarters and all 12 IHS areas and 9 EPA regions that administered the drinking water and clean water set-aside programs, discussing the data and any of its limitations. We tested the data for accuracy and completeness by identifying any duplicate, missing, or invalid records and cross-referencing with relevant datasets. We determined that IHS's SDS project-level needs data and information from EPA's 2013 report were sufficiently reliable to provide descriptive information on tribes' needs for drinking water and wastewater infrastructure projects for this report.

Further, we reviewed documentation on the Home Inventory Tracking System (HITS)—IHS's database containing home-specific information that the agency also uses in administering the Sanitation Facilities Construction program. The information in HITS includes each home's geographic location and individual sanitation deficiency, and IHS officials said in February 2018 that the system contained a total of 405,986 homes. We also interviewed IHS headquarters and area officials about this system's contents, uses, and limitations, and we compared this information to the agency's implementation plan and other documentation for HITS. We identified issues with the information contained in the system related to its completeness (whether it contains the correct number of homes in light of its purpose) and related to the accuracy of homes identified as having no deficiency, as we discuss in the report.

These issues were sufficient for us to determine that the number of homes in the system was incomplete and that deficiency level information was not accurate for all homes in the system. As a result, we did not assess the reliability of other information in HITS that was not relevant to our review. We also interviewed officials from the other five agencies regarding any efforts to collect information on tribal drinking water and wastewater infrastructure needs.

To determine the extent to which the agencies funded tribal drinking water and wastewater infrastructure projects, we analyzed data from the seven agencies administering programs that provide assistance to tribes for drinking water and wastewater infrastructure—IHS, EPA, USDA, HUD, Reclamation, EDA, and Corps. Specifically, we obtained and analyzed obligations data for drinking water and wastewater projects under programs that are specifically for or available to tribes. Generally, we reviewed each agency's obligations data for fiscal years 2012 through 2016, the most recent 5 years of data available at the time of our review. Corps provided us with information on obligations for projects that involved tribal drinking water or wastewater infrastructure, but none of these obligations were in fiscal years 2012 through 2016. We assessed the reliability of the other agencies' data by reviewing our previous related work regarding the use of these data and any available documentation from each agency; interviewing knowledgeable agency officials involved with collecting or analyzing these data; and testing data for accuracy and completeness by identifying any duplicate, missing, or invalid records. We present more details about each agency's data, any limitations, and how we addressed those limitations below. On the basis of these efforts, we determined that the data obtained from these agencies were sufficiently reliable for our descriptive purposes unless otherwise noted below.

- IHS. IHS provided us with project-level obligations data from fiscal years 2012 through 2016 for tribal drinking water and wastewater infrastructure projects from its Project Data System. In reviewing these data, we found data reliability issues that posed challenges to accurately reporting IHS's project obligations separate from other agencies' contributions to projects, which IHS also tracks in the system. We determined that the project-level obligations data from the Project Data System were not sufficiently reliable for the purposes of this objective. However, we determined that using IHS's information on allocations to areas for the same time frame would introduce fewer limitations to our reporting. IHS provided us with information from fiscal years 2012 through 2016 on allocations to each of its 12 areas by Sanitation Facilities Program activity (i.e., sanitation deficiencies, new housing, and emergency and special projects). IHS officials stated that the IHS Director of the Division of Sanitation Facilities Construction determines the area allocations amounts annually, and that IHS obligated all of its area allocations each fiscal year. IHS did not separate the area allocations information by drinking water, wastewater, or solid waste projects; therefore, we report total obligations with solid waste projects included.
- EPA. EPA provided us with project-level obligations data from fiscal years 2012 through 2016 from each of its three tribal-specific programs listed in Table 4. EPA uses its Tribal Direct Implementation Nexus system to track project obligations for the Clean Water Indian Set-Aside and Drinking Water Infrastructure Grants Tribal Set-Aside programs, but the agency relies on the State of Alaska to provide similar project-level information for its Alaska Native Villages and Rural Community Water Grant program. In reviewing EPA's data, we found several duplicate project records. We confirmed the issue with EPA officials and deleted those duplicate records to accurately aggregate EPA's obligations by fiscal year for our report.
- USDA. USDA provided us with grant and loan obligations data from fiscal years 2012 through 2016 for all of its programs specifically for or available to tribes from its Community Program Application Processing system. First, we removed solid waste and landfill projects that were indicated as such in the project name. To determine the project obligations for programs specifically for tribes, we included all obligations from USDA's Native American and Rural Alaska

Village Grant programs. USDA also awarded grants and loans to tribes or non-profit organizations working on behalf of tribes from non-tribal specific programs such as from its Water and Waste Disposal program as well as the Section 306C Colonias, Emergency Community Water Assistance Grant, Predevelopment Planning Grants, Special Evaluation Assistance for Rural Communities and Households, and Technical Assistance and Training programs. To determine the project obligations for those programs, we included projects that had an applicant or customer type as a tribe or tribal entity (e.g., an organization working on behalf of a tribe or tribes such as tribal health consortia or tribal utility authorities) and projects that served a population of at least 50 percent tribal users. For these awards, we included the full amount of the award regardless of the percent of tribal users served.

- HUD. HUD provided us with project-level obligations data from fiscal years 2012 through 2016 for its Indian Community Development Block Grant program from its Performance Tracking Database. We worked with HUD officials to identify projects that included drinking water and wastewater infrastructure and to identify the amount of the obligations used for those purposes to determine HUD's overall fiscal year project obligations for tribal water infrastructure.
- Reclamation. Reclamation provided us with project-level obligations data from fiscal years 2012 through 2016 for the tribal portions of authorized water system projects, including projects authorized by enacted Indian water rights settlements. For the Indian water rights settlement project obligations, Reclamation provided both mandatory and discretionary amounts. We included both rural water system projects and Indian water rights settlements projects in reporting Reclamation's overall fiscal year obligations.
- EDA. EDA provided us with project-level obligations data from fiscal years 2012 through 2016 for tribal projects funded by its Public Works, Economic Adjustment Assistance, and Planning programs from its Operations Planning and Control System. To determine whether the EDA projects included drinking water or wastewater infrastructure, we reviewed each project's description or scope of work for mention of a drinking water or wastewater infrastructure component. If we determined that the project included water infrastructure, we included the entire project's obligation amount for each fiscal year we report.

In addition, to determine the extent to which agencies' funding addressed the most severe sanitation deficiencies, we identified programs that have documented goals in regulation and policy to fund projects that meet these needs, which the programs identify as the absence of safe drinking water or wastewater disposal facilities. These selected programs included IHS's Sanitation Facilities Construction program, EPA's clean water set- aside program, and USDA's Native American program. For these programs, we compared the number of funded projects to address the most severe sanitation deficiencies with the number of funded projects that met other needs for fiscal year 2016. Specifically, for IHS and EPA, we calculated the percentage of projects for each deficiency level that the agencies and other entities selected to fund from the fiscal year 2016 SDS list. For USDA, we reviewed the list of Native American program project obligations in fiscal year 2016 and determined the number of projects where USDA reported the purpose as new, replacement, renovation, or expansion. We also reviewed documentation of the agencies' project identification and selection methods to determine whether these methods aligned with stated goals. We interviewed IHS and EPA officials from headquarters and all area and regional offices that administer these programs, and USDA officials from headquarters and six state-level offices (see below for state selection information), regarding their administration of these programs. Additionally, we analyzed IHS's data from the SDS from fiscal years 2005 through 2016 to identify projects that remained unfunded and that were in the SDS for more than 5 years. We did not review the extent to which EPA's drinking water set- aside program addressed the most severe sanitation deficiencies because EPA regions implement the program using a variety of different processes.

During the course of evaluating the extent to which federal agencies have provided funding for tribal drinking water and wastewater infrastructure projects, we identified issues with USDA's Rural Alaska Village Grant program. We reviewed obligations data in light of the program's authorizing statute, implementing regulations, and relevant provisions in USDA appropriations acts. USDA provided us with the Rural Alaska Village Grant program's award amounts for fiscal years 1997 through 2016, and we determined whether the grant recipients were eligible or ineligible at the time of the award. We interviewed agency officials who manage the program and from USDA's Office of the General Counsel.

To determine the extent to which the federal agencies collaborated to meet tribal water needs, we reviewed documentation of national-level collaboration, including federal program and interagency documents, such as national-level memorandums of understanding and interagency agreements. We interviewed headquarters officials from the seven agencies about their interagency collaboration. We compared the agencies' actions to the key features of interagency collaboration that we have previously identified.[108] We reviewed agencies' collaboration at the regional level by surveying the seven agencies about their joint actions on activities related to tribal drinking water and wastewater in six states— Alaska, Arizona, California, New York, Oklahoma, and South Dakota— and by conducting a network analysis using the survey responses. We selected agency regional offices within these six states as the unit of analysis because the federal

---

[108] GAO, *Managing for Results: Key Considerations for Implementing Interagency Collaborative Mechanisms*, GAO-12-1022 (Washington, D.C.: Sept. 27, 2012). Key features fall into the following categories: outcomes and accountability, bridging organizational cultures, leadership, clarity of roles and responsibilities, participants, resources, and written guidance and agreements.

agencies organize their field structures differently, with some using region, district, area, or state offices to work with tribes—we refer to all of these office types as regional offices. We selected the nonprobability sample of six states to include a large percentage of the number of federally recognized tribes, to obtain a range in the total federal obligations to tribes and identified needs of tribes in the SDS, and for geographic diversity. The sample of states is not generalizable, and the results of our work do not apply to all states where Indian tribes are located. However, reviewing federal agency collaboration in these states provides illustrative examples of interagency collaboration within the six selected states, which include about 70 percent of the 573 federally recognized tribes. We compared the agencies' reported collaboration with a national-level memorandum of understanding that contained commitments for collaborating at the regional level. For a detailed description of our survey methodology and the analysis of our results, see appendix II.

We also interviewed federal agency and State of Alaska officials to discuss the extent to which their drinking water and wastewater assistance programs collaborate with other agencies to meet tribal needs in the six selected states. We interviewed, either in person or by telephone, officials from the eight IHS areas, five EPA regions, and six USDA state offices that work with tribes and other agencies in the six states. We conducted site visits from February through April 2017 to three of the six states—Alaska, Arizona, and Oklahoma. During these visits, we met with tribal officials and staff and federal agency officials, and we visited tribal water infrastructure project sites. We selected these states for site visits based on geographic diversity and to obtain a range in the amount of tribal water infrastructure needs identified in the SDS. We met with or interviewed by telephone officials from 22 Indian tribes and representatives from 8 intertribal organizations that represent and work with tribes on water infrastructure issues to obtain their views about the water and wastewater infrastructure assistance that they receive from federal agencies. We judgmentally selected these tribes and organizations to obtain a range in their geographic locations and the amount and variety of federal drinking water and wastewater infrastructure assistance they have received. Our findings are not generalizable to all tribes but provide illustrative examples of input provided by tribal officials.

We conducted this performance audit from August 2016 to May 2018 in accordance with generally accepted government auditing standards. Those standards require that we plan and perform the audit to obtain sufficient, appropriate evidence to provide a reasonable basis for our findings and conclusions based on our audit objectives. We believe that the evidence obtained provides a reasonable basis for our findings and conclusions based on our audit objectives.

APPENDIX II: GAO SURVEY OF FEDERAL AGENCY COLLABORATION ON TRIBAL WATER INFRASTRUCTURE PROJECTS IN SIX SELECTED STATES

This appendix describes how we selected the sample and administered the survey, designed the survey questionnaire, and conducted the network analysis for our survey on interagency collaboration regarding tribal drinking water and wastewater infrastructure projects.

Sample Selection and Survey Administration

To determine the extent to which the selected federal agencies have collaborated to meet tribal water needs, we surveyed officials at seven federal agencies: Indian Health Service (IHS), Environmental Protection Agency (EPA), U.S. Department of Agriculture (USDA), Department of Housing and Urban Development (HUD), Economic Development Administration (EDA), Bureau of Reclamation, and the U.S. Army Corps of Engineers. Specifically, we surveyed agency officials in six states: Alaska, Arizona, California, New York, Oklahoma, and South Dakota.

Appendix I describes how we selected these agencies and states. The results of this survey are not generalizable beyond these agencies in these states.

We reviewed maps of each agency's regional or state offices and identified and confirmed the offices that work with tribes and other agencies in the six selected states. If one state included multiple regions from the same agency, we administered the survey to officials in all relevant regional offices. In addition, if one agency's region covered more than one of the selected states, we administered a survey to the agency's regional office for each state. The federal agencies and regional offices we included in our survey were:

- Corps divisions: Great Lakes & Ohio River, Northwestern, Pacific Ocean, South Pacific, Southwestern;
- EDA regions: Austin, Denver, Philadelphia, Seattle;
- EPA regions: 2, 6, 8, 9, 10 (Alaska Operations Office);
- HUD regions: Alaska, Eastern Woodlands, Northern Plains, Southern Plains, Southwest;
- IHS areas: Alaska, California, Great Plains, Nashville, Navajo, Oklahoma City, Phoenix, Tucson;
- Reclamation regions: Great Plains, Lower Colorado, Mid-Pacific; and
- USDA state offices: Alaska, Arizona, California, New York, Oklahoma, South Dakota.

In Alaska, we also included the Alaska Department of Environmental Conservation as a respondent because the state provides a 25 percent match for two federal water infrastructure programs. We did not include other state agencies because they do not provide a similar match. We also included the Alaska Native Tribal Health Consortium because it administers IHS's Sanitation Facilities Construction program in Alaska.

The practical difficulties of conducting any survey may introduce errors, commonly referred to as non-sampling errors. For example, respondents may have difficulty interpreting a question, they may have limited information to respond to a question, or officials from different agencies may have different recollections regarding the extent of collaboration on a particular project. We sought to minimize the impact of non-sampling error by conducting six pretests of the draft questionnaire with agency officials; five pretests were conducted by telephone and one pretest was conducted in person. We selected officials to cover a range of agencies and locations. During these pretests, we sought to determine whether (1) the questions were clear and unambiguous, (2) terminology was used correctly, (3) the questionnaire did not place an undue burden on agency officials, (4) the information could feasibly be obtained, and (5) the survey was comprehensive and unbiased. We modified the questionnaire in response to these pretests. To further minimize the impact of non- sampling error, we conducted a sensitivity analysis.

We customized the questionnaire for each agency regional office so that we asked each office to respond about its collaboration only with the other agencies located in its state. We e-mailed these questionnaires to 46 respondents from May 15 through May 17, 2017, and conducted follow-up as necessary. We received a 100 percent response rate.

Survey Questionnaire Design

In the survey, we asked each agency regional office whether it had jointly conducted activities related to tribal drinking water or wastewater projects during the past 3 years with each of the other agencies' regional offices within the same state. If an agency regional office responded "yes," we then provided a list of tribal drinking water and wastewater activities and asked the agency regional office if it had jointly conducted any of the listed activities related to tribal drinking water and wastewater infrastructure projects in collaboration with the other agency. The activities included: identifying infrastructure needs, communicating information to tribes about programs that fund projects, planning and designing proposed projects, evaluating proposed projects according to eligibility and scoring criteria, selecting projects to fund, constructing projects, providing technical assistance for operating and maintaining water infrastructure, and negotiating or implementing Indian water rights settlements.[109] We developed the list of activities based on our initial interviews and pretests with agency officials.

We next provided a list of collaborative mechanisms.[110] For each of these collaborative mechanisms, we asked the agency regional office if it had used the mechanism when jointly conducting activities in collaboration with the other agency related to tribal drinking water and wastewater infrastructure projects within the same state during the past 3 years. The mechanisms included: state-, regional-, or project-level memorandum of understanding or agreement; interagency agreement to transfer funding; working group, task force, or committee; consulting on project selection; sharing project documents; geographic co-location; shared database or other data sharing; conferences or forums; informal or ad hoc communication; and personnel detailing or sharing. If the agency regional office responded that it had not used one of the listed mechanisms, we asked if it would be beneficial to use that mechanism to collaborate in the future. We identified the list of mechanisms based on our prior work on interagency collaboration and pretests with agency officials.[111] We also asked the agency regional office what factors, if any, helped it to collaborate with the other agency on tribal drinking water and wastewater infrastructure projects in the state and what factors, if any, hindered it from collaborating with the other agency. For both questions, we asked the agency regional office to consider agency policies and procedures, available resources, leadership, personalities, presence of written agreements, and accountability measures.[112]

If an agency regional office responded "no" to the initial question of whether it had jointly conducted activities related to tribal drinking water or wastewater projects during the past 3 years with another agency's regional office, we asked a shorter set of follow-up questions. We provided the list of collaborative mechanisms and asked if it would be beneficial for the agency regional office to use any of the listed mechanisms to collaborate with the other agency on activities related to tribal drinking water and wastewater infrastructure projects in the future in the state. We also asked the agency regional office to describe the factors, if any, that hindered its collaboration with the other agency.

---

[109] We did not include negotiating or implementing Indian water rights settlements in the list of wastewater activities based on input we received during pretesting.

[110] We used the term "method" instead of "mechanism" in the survey to improve respondents' understanding based on input we received during pretesting.

[111] GAO, *Managing for Results: Key Considerations for Implementing Interagency Collaborative Mechanisms*, GAO-12-1022 (Washington, D.C.: Sept. 27, 2012).

[112] We created this list of items to consider based on our prior work on interagency collaboration and pretests with agency officials. See GAO-12-1022.

NETWORK ANALYSIS

To quantify the extent of interagency collaboration during the past 3 years and the potential for future collaboration among the federal agencies we surveyed, we conducted a Network Analysis—a method of analyzing the patterns of interaction among multiple entities. Specifically, we aggregated the survey responses to our questions about drinking water and wastewater activities and collaborative mechanisms for each pair of agencies in all six states.[113] We configured these aggregated data into networks representing the pattern of collaboration among the agencies. We then analyzed these networks to determine how extensively the agencies have collaborated and the extent to which additional future collaboration could be beneficial for them. We also analyzed these networks to assess how the pattern of collaboration varied by state. We describe the steps of our analysis and agency survey responses below.

Quantifying Collaboration between Pairs of Federal Agencies

To quantify the extent of collaboration among the federal agencies across the six states during the past 3 years, we aggregated the responses to our survey by agency pair. The seven federal agencies form 21 possible agency pairs. For each agency pair, we combined the first agency's responses regarding its collaboration with the second agency and the second agency's responses regarding its collaboration with the first agency.[114] We aggregated the agency pair responses in this way for each of the three measures of collaboration for all six states, specifically:

- Drinking water and wastewater activities. We calculated the total number of instances in which each agency in a pair reported having worked on an activity with the other agency in that pair during the past 3 years (see column 2 in Table 5). We examined this measure to identify the pairs of agencies that collaborated most and least extensively. For example, IHS and EPA reported the highest number of instances of jointly conducting tribal drinking water and wastewater

[113] In Alaska, Arizona, and California, for agencies where we surveyed more than one of their regional offices, we combined the offices' answers to conduct the network analysis. Specifically, we combined responses from: IHS's Alaska Area and the Alaska Native Tribal Health Consortium (Alaska); IHS's Phoenix, Tucson, and Navajo areas (Arizona); IHS's California and Phoenix areas (California); and Reclamation's Mid-Pacific and Lower Colorado regions (California).

[114] To assess the robustness of the analysis presented here, we also separately conducted a sensitivity analysis using only mutual collaboration between agency pairs, and the key findings from our network analysis remained the same. In particular, in our sensitivity analysis for each of the three measures of collaboration (drinking water and wastewater activities, use of collaborative mechanisms, and potential future collaboration), we defined collaboration as existing only if both agencies in a pair reported collaborating with the other. If so, then the value of the link between the two agencies became the sum of all instances of each agency in a pair reporting working together with the other agency. While the pair-specific collaboration values differed to some degree under the new assumptions, the key network analysis results presented in this appendix still held.

activities across the six states. In contrast, EDA and IHS reported no such instances of collaboration.

Table 5. Federal agency collaboration and potential for future collaboration on tribal water infrastructure activities

Agency pair	Instances of agencies reporting having jointly conducted an activity	Instances of agencies reporting having used a mechanism to collaborate	Instances of agencies reporting it would be beneficial to use an additional mechanism to collaborate in the future
EPA – IHS	153	96	9
IHS – USDA	119	60	38
EPA – USDA	50	28	39
HUD – IHS	32	31	57
IHS – Reclamation	32	36	38
EDA – USDA	26	22	31
HUD – USDA	25	19	40
Corps – HUD	23	23	43
EPA – HUD	16	20	48
Corps – USDA	14	12	47
Reclamation – USDA	13	14	28
Corps – Reclamation	12	10	29
Corps – EPA	6	6	49
Corps – EDA	5	6	44
Corps – IHS	5	6	69
EDA – EPA	4	4	42
EPA – Reclamation	3	7	23
HUD – Reclamation	1	3	31
EDA – HUD	0	0	53
EDA – IHS	0	0	60
EDA – Reclamation	0	0	28
Total	539	403	846

Source: GAO analysis of survey responses. | GAO-18-309.

Notes: Agencies include Bureau of Reclamation, Department of Housing and Urban Development (HUD), Economic Development Administration (EDA), Environmental Protection Agency (EPA), Indian Health Service (IHS), U.S. Army Corps of Engineers, and U.S. Department of Agriculture (USDA) in the states of Alaska, Arizona, California, New York, Oklahoma, and South Dakota. The time frame covered by this table is approximately May 2014 through April 2017, regarding the agencies reporting having jointly conducted an activity and having used a mechanism to collaborate. We disseminated the survey in May 2017 and asked agencies to report about their collaboration during the 3-year period prior to our survey.

- Use of collaborative mechanisms. We calculated the total number of instances in which each agency in a pair reported having used a mechanism to collaborate with the other agency in that pair during the past 3 years (see column 3 in Table 5). We examined this measure to identify the pairs of agencies that collaborated most and least extensively. The pattern of collaboration based on this measure is similar to the pattern based on drinking water and wastewater activities. For example, IHS and EPA also reported the highest number of instances of using specific collaborative mechanisms across the six states.

- Potential future collaboration. We calculated the total number of instances in which each agency in a pair reported that it would be beneficial to use a mechanism to collaborate with the other agency in that pair in the future (see column 4 in Table 5). We compared this measure to the number of mechanisms the agency pairs reported having used during the past 3 years. Each of the agency pairs reported that it would be beneficial to use additional collaborative mechanisms in the future, including those pairs that had reported not collaborating. For example, the agency pairs of EDA-IHS and EDA-Reclamation both reported no instances of using a mechanism to collaborate with each other and both reported multiple instances in which use of a collaborative mechanism would be beneficial in the future.[115]

Quantifying the Potential for Increased Collaboration

To quantify the potential to increase collaboration among the federal agencies, we configured the agency pair data into two networks. The first network represented recent collaboration among the agencies—the instances in which agencies reported having used a mechanism to collaborate during the past 3 years (based on column 3 in Table 5).

The second network represented potential future collaboration among the agencies (based on the sum of columns 3 and 4 in Table 5). As such, it captures the instances in which agencies reported having used a mechanism to collaborate during the past 3 years plus the instances in which they reported it would be beneficial to use an additional mechanism in the future.

Figure 3 shows a graphical illustration of these two networks. In this figure, the circles represent agencies and the lines represent collaboration between the agencies. Specifically, the darkness of the lines indicates the number of mechanisms used by the corresponding pair of agencies. The left side of Figure 3 illustrates reported use of collaborative mechanisms during the past 3 years, and the right side of Figure 3 illustrates potential future collaboration. The figure shows that overall collaboration would increase if the agencies began using the additional mechanisms that they reported would be beneficial.

[115] The EDA-IHS agency pair reported 60 instances of potentially beneficial future collaboration—the second highest among our agency pairs—while the EDA-Reclamation agency pair reported 28 instances—the second lowest among our agency pairs.

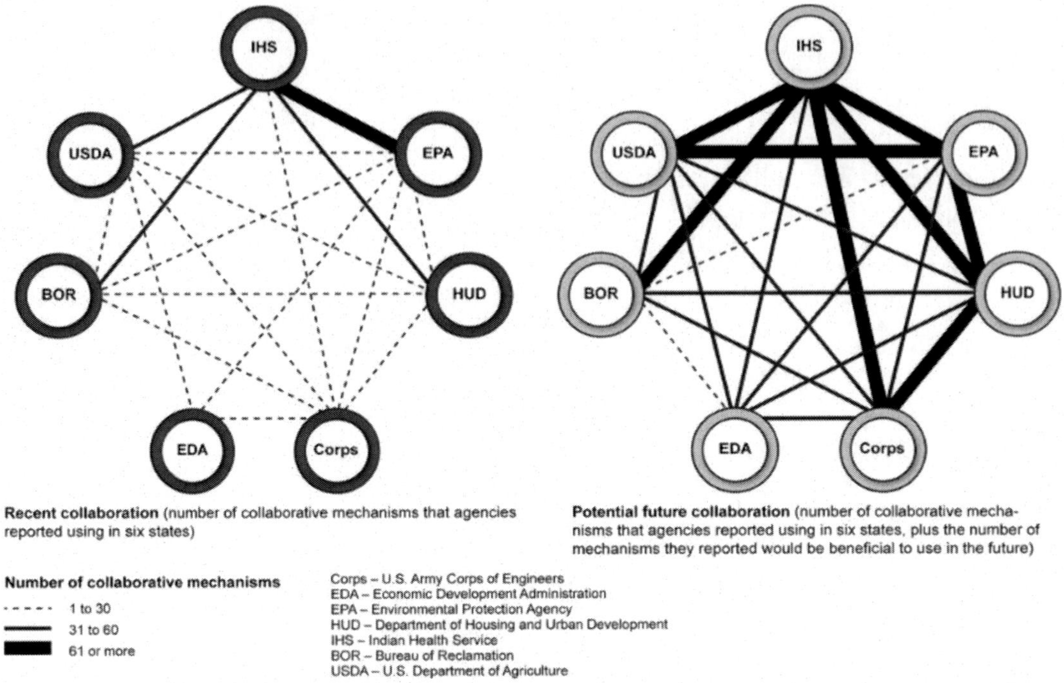

Source: Analysis of GAO survey of seven federal agencies in six states. | GAO-18-309.

Notes: A count of one collaborative mechanism refers to a single agency in a single state reporting that it used a mechanism to collaborate with a second agency in that state or that it would be beneficial to do so in the future. We surveyed agency regional offices in Alaska, Arizona, California, New York, Oklahoma, and South Dakota. The time frame covered by this figure for recent collaboration is approximately May 2014 through April 2017. We disseminated the survey in May 2017 and asked agencies to report about their collaboration during the 3-year period prior to our survey.

Figure 3. Federal Agency Collaboration and Potential for Future Collaboration on Tribal Drinking Water and Wastewater Infrastructure Activities.

We quantified the difference between these networks in two ways. First, we calculated the increase in overall collaboration that would occur if agencies began using the additional mechanisms that they reported would be beneficial. Based on this calculation, the number of instances of agencies using collaborative mechanisms would approximately triple.

Specifically, agencies reported 403 instances of having used a specific mechanism to collaborate with another agency—this number would increase to 1,249 if agencies began using all of the identified mechanisms that they reported would be beneficial. This difference is shown in Figure 3, in which the right side of the figure (potential future collaboration) has a greater number of darker lines connecting the agencies compared with the left side of the figure (recent collaboration).

Second, we measured how the relative amount of collaboration for each agency would change if the agencies began using additional mechanisms they reported would be beneficial. To do this, we aggregated the agency pair data for each of the agencies. For the network of recent collaboration, for example, we added (1) the number of instances that each agency reported using a collaborative mechanism with any of the other agencies and (2) the number of instances that any of the other agencies reported using a collaborative mechanism with the first agency. We performed a similar calculation using the agency pair data for the network of potential future collaboration. The analysis shows that the use of collaborative mechanisms during the past 3 years was primarily centered on three agencies (IHS, EPA, and USDA). If all of the agencies began using the additional mechanisms that they reported would be beneficial, however, collaboration would be distributed more evenly across the entire network of agencies. This difference is also shown in Figure 3, in which agencies such as HUD, Reclamation, and Corps are connected to other agencies with dashed lines on the left side of the figure (representing less extensive recent collaboration), but with thick lines on the right side of the figure (representing more extensive potential future collaboration).

Quantifying the Variation in Collaboration by State

Table 6. Agency collaboration on tribal drinking water and waste water infrastructure activities

	State						
Tribal drinking water and wastewater infrastructure activity	Alaska	Arizona	California	New York	Oklahoma	South Dakota	Total
Identifying water infrastructure needs	33	23	24	5	12	20	117
Communicating information to tribes about programs that fund projects	36	26	29	5	12	24	132
Planning and designing proposed projects	27	13	13	3	10	19	85
Evaluating proposed projects according to eligibility and scoring criteria	25	10	10	4	6	15	70
Selecting projects to fund	26	11	11	4	6	13	71
Constructing projects	28	9	11	4	11	12	75
Negotiating or implementing Indian water rights settlements	0	0	0	0	0	0	0
Providing technical assistance for operating and maintaining infrastructure	24	12	10	3	8	21	78
Total	199	104	108	28	65	124	628

Source: GAO analysis of survey responses. | GAO-18-309.

Notes: A single instance in these data represents one agency reporting it jointly conducted a specific activity with another agency in a specific state. This table shows agencies' combined responses about drinking water and wastewater infrastructure activities except for negotiating or implementing Indian water rights settlements, which we only asked about with respect to drinking water infrastructure. The agencies were: Bureau of Reclamation, Department of Housing and Urban Development, Economic Development Administration, Environmental Protection Agency, Indian Health Service, U.S. Army Corps of Engineers, U.S. Department of Agriculture, and the Alaska Department of Environmental Conservation, in the states of Alaska, Arizona, California, New York, Oklahoma, and South Dakota. The time frame covered by this table is approximately May 2014 through April 2017. We disseminated the survey in May 2017 and asked agencies to report about their collaboration during the 3-year period prior to our survey.

Table 7. Agency collaboration on tribal water infrastructure activities using specific mechanisms

Collaborative mechanism	State						
	Alaska	Arizona	California	New York	Oklahoma	South Dakota	Total
State-, regional-, or project-level Memorandum of Understanding or Agreement	12	3	4	1	1	7	28
Interagency Agreement to transfer funding	8	2	4	2	3	4	23
Working group/task force/committee (formal or informal)	20	14	16	0	7	18	75
Consulting on project selection	13	12	10	1	3	15	54
Sharing of project documents (e.g., preliminary engineering report, project summary, environmental analysis)	17	13	11	2	5	15	63
Geographic co-location (office sharing)	1	1	1	0	0	2	5
Shared database or other data sharing	13	11	9	2	3	15	53
Conferences/forums	16	12	14	2	7	17	68
Informal or ad hoc communications	20	14	15	2	6	19	76
Personnel detailing or sharing	1	3	3	1	1	1	10
Total	121	85	87	13	36	113	455

Source: GAO analysis of survey responses. | GAO-18-309

Notes: A single instance in these data represents one agency reporting it worked together with another agency using a specific collaborative mechanism in a specific state. The agencies were: Bureau of Reclamation, Department of Housing and Urban Development, Economic Development Administration, Environmental Protection Agency, Indian Health Service, U.S. Army Corps of Engineers, U.S. Department of Agriculture, and the Alaska Department of Environmental Conservation, in the states of Alaska, Arizona, California, New York, Oklahoma, and South Dakota. The time frame covered by this table is approximately May 2014 through April 2017. We disseminated the survey in May 2017 and asked agencies to report about their collaboration during the 3-year period prior to our survey.

To quantify the extent of variation in collaboration by state, we disaggregated the agency pair data reported in Table 5 by each of the states for the three measures of collaboration we asked about in our survey. In particular, Tables 6, 7, and 8 show the number of instances in which an agency reported collaborating on drinking water and wastewater infrastructure activities with another agency during the past 3 years (Table 6), using collaborative mechanisms with another agency during the past 3 years (Table 7), and collaborative mechanisms that would be beneficial to use with another agency in the future (Table 8). The totals in the bottom rows of these tables show the extent of collaboration based on these measures by state. Specifically, Tables 6 and 7 show that agencies worked together on activities and used collaborative mechanisms most extensively in Alaska and least extensively in New York and Oklahoma.

Table 8 shows that agencies in New York and Oklahoma reported the greatest potential for using additional collaborative mechanisms. The totals in the far right columns of these tables show the extent of reported collaboration by activity (Table 6), collaborative mechanism (Table 7), and the extent of potential future collaboration by collaborative mechanism (Table 8).

Table 8. Potential future agency collaboration on tribal water infrastructure activities using specific mechanisms

Collaborative mechanism	**State**						
	Alaska	Arizona	California	New York	Oklahoma	South Dakota	Total
State-, regional-, or project-level Memorandum of Understanding or Agreement	13	13	20	23	23	19	111
Interagency Agreement to transfer funding	4	10	9	11	10	2	46
Working group/task force/committee (formal or informal)	14	17	11	26	15	15	98
Consulting on project selection	11	14	13	17	28	4	87
Sharing of project documents (e.g., preliminary engineering report, project summary, environmental analysis)	16	26	20	24	31	18	135
Geographic co-location (office sharing)	1	1	7	0	2	1	12
Shared database or other data sharing	12	18	17	16	20	16	99
Conferences/forums	13	17	17	22	30	17	116
Informal or ad hoc communications	12	25	19	26	34	18	134
Personnel detailing or sharing	5	2	8	8	2	3	29
Total	102	143	141	173	195	113	867

Source: GAO analysis of survey responses. | GAO-18-309.

Note: A single instance in these data represents one agency reporting it would be beneficial to use a specific collaborative mechanism with another agency in a specific state in the future. The agencies were: Bureau of Reclamation, Department of Housing and Urban Development, Economic Development Administration, Environmental Protection Agency, Indian Health Service, U.S. Army Corps of Engineers, U.S. Department of Agriculture, and the Alaska Department of Environmental Conservation, in the states of Alaska, Arizona, California, New York, Oklahoma, and South Dakota.

Appendix III: Federal Agency Obligations for Tribal Drinking Water and Wastewater Infrastructure Projects, Fiscal Years 2012 through 2016

Table 9. Federal agency obligations for tribal drinking water and wastewater infrastructure projects, fiscal years 2012 through 2016 (Dollars in millions)

		Fiscal year					
Agency and program		2012	2013	2014	2015	2016	Total obligations[a]
Indian Health Service[b]	New Housing	35.5	33.6	35.5	35.3	40.8	411.9
	Sanitation Deficiency	42.4	40.1	42.4	42.6	56.9	
	Special and Emergency	1.8	1.6	1.6	1.1	0.9	
Environmental Protection Agency[c]	Drinking Water Infrastructure Grants Tribal Set-Aside	18.0	17.4	17.6	19.5	18.9	277.6
	Clean Water Indian Set-Aside	28.9	26.9	28.5	26.7	30.0	
	Alaska Native Villages and Rural Communities	7.3	6.7	8.5	7.1	15.7	
U.S. Department of Agriculture[d]	Native American	15.0	11.1	13.5	24.7	22.7	278.2
	Rural Alaska Village Grant	29.4	27.7	21.5	21.6	32.0	
	Water and Waste Disposal grants	32.0	4.3	5.0	2.6	3.9	
	Other grants[e]	5.0	0.7	1.0	3.1	1.4	
	Water and Waste Disposal loans	19.2	4.8	3.3	3.7	6.1	
Bureau of Reclamation	Rural water system projects	24.1	26.8	34.8	32.8	38.7	567.8
	Indian water rights settlement provisions for rural water supply projects	64.4	76.8	77.1	88.7	103.7	
Housing and Urban Development	Indian Community Development Block Grant	6.3	2.0	3.1	5.1	4.5	21.0
Economic Development Administration[f]	Public Works; Economic Adjustment Assistance; Planning	0.0	1.8	0.2	0.1	0.3	2.4
Total		309.9	277.7	290.0	311.0	370.4	1,559.0

Source: GAO analysis of federal agency data. | GAO-18-309. (Continued …)

Notes: We included federal agencies' and programs' obligations to federally recognized tribes and tribal entities that were for or partially for drinking water and wastewater infrastructure. U.S. Army Corps of Engineers did not make obligations to tribes specifically for drinking water and wastewater infrastructure projects from fiscal years 2012 through 2016.

[a]Obligations totals do not include loans, and discrepancies in totals are due to rounding.

[b]The amount of Indian Health Service obligations are based on the agency's allocations to its 12 areas per fiscal year. Indian Health Service officials stated that all areas obligated their entire allocation each year. Indian Health Service information included projects to address drinking water, wastewater, and solid waste deficiencies.

[c]According to Environmental Protection Agency officials, obligations listed may not match annual appropriations because the agency may have de-obligated and re-obligated any unexpended obligations to other projects.

[d]We determined that the U.S. Department of Agriculture awarded a grant or loan from its non-tribal specific programs for a tribal drinking water or wastewater infrastructure project if the recipient was a tribe or tribal entity (for example, an organization working on behalf of a tribe or tribes such as tribal health consortia or tribal utility authorities) and if the project was to serve a population of at least 50 percent American Indian or Alaska Native.

[e]The U.S. Department of Agriculture also awarded grants to tribes or other tribal entities from non- tribal specific programs administered by headquarters. These programs include the Section 306C Colonias, Emergency Community Water Assistance Grant, Predevelopment Planning Grants, Special Evaluation Assistance for Rural Communities and Households, and Technical Assistance and Training programs.

[f]The Economic Development Administration obligated approximately $34,000 for one project in fiscal year 2012, which is not reflected in the table due to rounding. We determined that the Economic Development Administration awarded a grant for a tribal drinking water or wastewater infrastructure project if the project's description or scope of work mentioned a drinking water or wastewater infrastructure component. Obligations are combined from three programs: Public Works, Economic Adjustment Assistance, and Planning.

Appendix IV: Examples of Tribal Water Infrastructure Projects We Visited

This appendix contains summaries and photographs of selected tribal drinking water and wastewater infrastructure projects we visited from February through April 2017 in Alaska, Arizona, and Oklahoma.

Portable Alternative Sanitation System Pilot Project, Native Village of Kivalina, Alaska

The Native Village of Kivalina, located on a barrier island above the Arctic Circle, is one of approximately 30 communities in Alaska where residents do not have access to safe drinking water and wastewater disposal facilities in their homes. Kivalina, a community of 469 residents, has a community washeteria with washing machines, dryers, and drinking water available for purchase. Like many Alaska Native villages, the harsh winter climate, limited revenue, and isolation create challenges for installing and operating water infrastructure. Erosion due to diminishing sea ice and other factors threaten Kivalina, and the community is considering relocation. As such, infrastructure improvements are limited to small projects consisting of moveable, low-water use infrastructure to provide interim sanitation improvements. In 2015, the Alaska Native Tribal Health Consortium installed a pilot sanitation system in nine homes. This system is called the Portable Alternative Sanitation System and consists of a bathroom sink, rainwater catchment, in-home water treatment, and a separating toilet, where liquid waste is collected separate from solid waste. According to a Consortium report, the system is a low-cost alternative to traditional piped infrastructure.

Bathroom with separating toilet and sink. Rainwater catchment pipes on side of home. In-home storage tank for treating rainwater.

Source: GAO. | GAO-18-309.

Figure 4. Portable Alternative Sanitation System Pilot Project, Native Village of Kivalina, Alaska (April 2017).

The total cost was $633,000 to design, install, and monitor the system, with the Indian Health Service (IHS) and the Consortium contributing to the project. The Consortium recommended expanding the pilot system to the rest of Kivalina, and a Consortium official said it is working with IHS to test the system in several homes in three other unserved communities in Alaska.

Village of Shungopavi Sewer Line Q & Dump Stations Construction Project, Hopi Tribe, Arizona

As of 2015, more than 30 percent of the nearly 80 homes in the Hopi Village of Shungopavi did not have adequate wastewater disposal. The Sewer Line Q and Dump Stations construction project included installing a sewer main to connect nine homes to sewer service. Previously, some of these homes had discharged wastewater directly onto the ground, and one had a septic system. The project also involved installing three honeybucket dump stations in the village and connecting them to the existing sewer system so that an additional 19 homes could dispose of raw sewage in an environmentally safe manner. According to IHS officials, solid rock a few feet beneath the surface made it challenging and expensive to lay the sewer pipes. The total estimated cost was $666,000, with the Environmental Protection Agency (EPA), the Village of Shungopavi, and IHS contributing to the project. According to IHS officials, the project is expected to be fully constructed in 2018.

Digging a trench for sewer pipes.

Laying sewer pipes.

Source: GAO. | GAO-18-309.

Figure 5. Sewer Line Construction in the Village of Shungopavi, Hopi Tribe, Arizona (February 2017).

Oaks Wastewater Lagoons Construction Project, Cherokee Nation, Oklahoma

The Cherokee Nation's Oaks Wastewater Lagoons project serves an estimated 85 Indian-owned homes in the community of Oaks, Oklahoma. The project consisted of constructing three wastewater lagoons and a spray irrigation field. According to a tribal official, because the previous lagoons leaked into the adjacent creek, local residents who used the creek for swimming, fishing, and other traditional purposes were at high risk of coming in contact with lagoon leakage. The total cost of the project was an estimated $1.22 million, and the U.S. Department of Agriculture, EPA, IHS, the Department of Housing and Urban Development, and the Oklahoma Water Resources Board made contributions to the project. The Cherokee Nation completed the project in 2012 under the provisions of its self-governance compact with IHS.

Site of former wastewater lagoons with adjacent creek.

Newly constructed wastewater lagoons.

Source: GAO. | GAO-18-309.

Figure 6. Oaks Wastewater Lagoons and Former Lagoon Site, Cherokee Nation, Oklahoma (March 2017).

Drinking Water Pump Station Replacement Project, Sasakwa Rural Water District, Seminole Nation of Oklahoma

The Sasakwa Rural Water District is owned and operated by the Seminole Nation of Oklahoma and serves 61 households—about 60 percent of which are Indian homes, according to tribal officials. The Drinking Water Pump Station Replacement project involved drilling new wells and constructing a new pump station and treatment system. IHS constructed the original Sasakwa water treatment plant in 1972.

Exterior and interior of new pump station building, with operator work station.

Source: GAO. | GAO-18-309.

Figure 7. Drinking Water Pump Station Replacement Project, Sasakwa Rural Water District, Seminole Nation of Oklahoma (March 2017).

According to an IHS project summary, the problems with the prior system included (1) recurring leaks in the water transmission line and distribution system and (2) deterioration of the pump and treatment building and equipment due to weather, vandalism, and poor water quality. The project cost approximately $700,000, with EPA funding the project. According to tribal officials, the replacement water treatment plant became operational in 2014.

APPENDIX V: COMMENTS FROM THE DEPARTMENT OF HEALTH AND HUMAN SERVICES

DEPARTMENT OF HEALTH & HUMAN SERVICES   OFFICE OF THE SECRETARY

Assistant Secretary for Legislation
Washington, DC 20201

APR 1 1 2018

Anne-Marie Fennell and Alfredo Gomez
Director, Natural Resources and Environment
U.S. Government Accountability Office
441 G Street NW
Washington, DC 20548

Dear Ms. Fennell and Mr. Gomez:

Attached are comments on the U.S. Government Accountability Office's (GAO) report entitled, "*Drinking Water and Wastewater Infrastructure: Opportunities Exist to Enhance Federal Agency Needs Assessment and Coordination on Tribal Projects*" (GAO-18-309).

The Department appreciates the opportunity to review this report prior to publication.

Sincerely,

Matthew D. Bassett
Assistant Secretary for Legislation

GENERAL COMMENTS FROM THE DEPARTMENT OF HEALTH & HUMAN SERVICES ON THE GOVERNMENT ACCOUNTABILITY OFFICE'S DRAFT REPORT ENTITLED - DRINKING WATER AND WASTEWATER INFRASTRUCTURE: OPPORTUNITIES EXIST TO ENHANCE FEDERAL AGENCY NEEDS ASSESSMENT AND COORDINATION ON TRIBAL PROJECTS (GAO-18-309)

The U.S. Department of Health & Human Services (HHS) appreciates the opportunity from the Government Accountability Office (GAO) to review and comment on this draft report.

Recommendation 1
The Director of the Indian Health Service (IHS) should implement a targeted, resource-efficient method to identify additional eligible Indian homes that may have existing deficiencies to include in Home Inventory Tracking System (HITS).

HHS Response
HHS concurs with GAO's recommendation.

IHS will continue to reach out to the Tribes to inform them of our joint role and desire to identify the additional eligible Indian homes that may have existing deficiencies.

Recommendation 2
The Director of IHS should implement a mechanism to indicate whether each home with a deficiency level of 0 has been assessed in HITS.

HHS Response
HHS concurs with GAO's recommendation.

IHS will create a new category for homes with a deficiency level 0 that have not been accessed and/or assessed.

Recommendation 3
The Director of IHS should reassess the point distribution across the Sanitation Deficiency System (SDS) scoring factors as part of its program guidelines update, in light of trade-offs between funding projects that address the most severe sanitation deficiencies and projects that meet other needs.

HHS Response
HHS concurs with GAO's recommendation.

As part of the update to the SDS Guidelines, IHS will reassess the scoring factors and submit updated SDS Guidelines to the Tribes for consultation.

Recommendation 7
The Director of IHS, in cooperation with other members of the tribal infrastructure task force, should review the 2011 task force report and identify and implement additional actions to help increase the task force's collaboration at the national level.

HHS Response
HHS concurs with GAO's recommendation.

GENERAL COMMENTS FROM THE DEPARTMENT OF HEALTH & HUMAN SERVICES ON THE GOVERNMENT ACCOUNTABILITY OFFICE'S DRAFT REPORT ENTITLED - DRINKING WATER AND WASTEWATER INFRASTRUCTURE: OPPORTUNITIES EXIST TO ENHANCE FEDERAL AGENCY NEEDS ASSESSMENT AND COORDINATION ON TRIBAL PROJECTS (GAO-18-309)

IHS will re-engage with Infrastructure Task Force (ITF) members on recommended streamlining opportunities from the February 2011 report titled, "Overview of Tribal Infrastructure Funding Application Processes and Recommended Streamlining Opportunities."

Recommendation 12
The Director of IHS, in cooperation with other members of the tribal infrastructure task force, should direct IHS area offices to identify and pursue additional mechanisms to increase their collaboration.

HHS Response
HHS concurs with GAO's recommendation.

IHS will direct IHS Area Offices to identify and pursue additional mechanisms to increase their collaboration with regional ITF representatives and other technical assistance providers.

APPENDIX VI: COMMENTS FROM THE DEPARTMENT OF THE INTERIOR

United States Department of the Interior
OFFICE OF THE SECRETARY
Washington, DC 20240

MAR 2 3 2018

Ms. Anne-Marie Fennell
Director, Natural Resources and Environment
U.S. Government Accountability Office
441 G Street, NW
Washington, DC 20548

Dear Ms. Fennell:

Thank you for providing the Department of the Interior (Department) the opportunity to review and comment on the draft Government Accountability Office (GAO) report entitled, *Drinking Water and Wastewater Infrastructure: Opportunities Exist to Enhance Federal Agency Needs Assessment and Coordination on Tribal Projects* (GAO-18-309). We appreciate GAO's review of Federal programs that provide funding for drinking water and wastewater infrastructure.

The Department notes the following GAO recommendations:

Recommendation 11: The Commissioner of Reclamation, in cooperation with other members of the tribal infrastructure task force, should review the 2011 task force report and identify and implement additional actions to help increase the task force's collaboration at the national level.

Response: Concur. Reclamation will meet with the infrastructure task force to review the 2011 task force report and identify additional actions that increase the task force's collaboration at the national level. Reclamation continues to take steps to improve collaboration with other agencies by attending meetings and conferences where tribes and agencies discuss this and similar issues.

Recommendation 16: The Commissioner of Reclamation, in cooperation with other members of the tribal infrastructure task force, should direct Reclamation regional offices to identify and pursue additional mechanisms to increase their collaboration.

Response: Concur. Reclamation has worked with task force member agencies and some Reclamation regional officials to encourage an increased level of collaboration with tribes and other agencies. Further direction and support will be made as part of on-going Reclamation management planning and implementation efforts.

Sincerely,

Timothy R. Petty, Ph.D.
Assistant Secretary
for Water and Science

APPENDIX VII: COMMENTS FROM THE ENVIRONMENTAL PROTECTION AGENCY

UNITED STATES ENVIRONMENTAL PROTECTION AGENCY
WASHINGTON, D.C. 20460

APR – 6 2018

OFFICE OF WATER

Mr. J. Alfredo Gomez, Director
Ms. Anne-Marie Fennell, Director
Natural Resources and Environment
U.S. Government Accountability Office
Washington, D.C. 20548

Dear Mr. Gomez and Ms. Fennell:

Thank you for the opportunity to review and comment on the Government Accountability Office's (GAO) draft report, "Drinking Water and Wastewater Infrastructure: Opportunities Exist to Enhance Federal Agency Needs Assessment and Coordination on Tribal Projects" (GAO-18-309).

The purpose of this letter is to provide the U.S. Environmental Protection Agency's (EPA) response to the draft report findings, conclusions, and recommendations. In this report, the GAO examined the extent to which federal agencies: (1) identify Indian tribes' drinking water and wastewater infrastructure needs, (2) fund tribal drinking water and wastewater infrastructure projects to address the most severe sanitation deficiencies, and (3) collaborate to meet Indian tribes' drinking water and wastewater infrastructure needs.

The GAO found that tribal infrastructure programs could benefit from improved collaboration. The report referenced that several agencies, including the EPA, reported examples of duplicative efforts that could be eliminated through better communications amongst federal partners at the regional level. The GAO also suggested that implementation of recommendations from a 2011 report prepared for the multi-agency tribal Infrastructure Task Force (ITF) could improve ITF members' abilities to leverage limited federal resources. The EPA agrees with the GAO's findings, conclusions, and recommendations relevant to the Agency, with minor suggested edits noted below.

GAO Recommendations:

Two of the sixteen recommendations in the report were directed at the EPA:

- The Administrator of EPA, in cooperation with other members of the tribal infrastructure task force, should review the 2011 task force report and identify and implement additional actions to help increase the task force's collaboration at the national level. (Recommendation 8)

- The Administrator of EPA, in cooperation with other members of the tribal infrastructure task force, should direct EPA regional offices to identify and pursue additional mechanisms to increase their collaboration. (Recommendation 14)

EPA Response:

The EPA agrees with the recommendation to review the 2011 report prepared by the Streamlining Preconstruction Paperwork Workgroup for the ITF, and to identify and implement additional actions to help increase collaboration at the national level. The EPA has reached out to our federal partners and agreed to dedicate a portion of the next ITF meeting (scheduled for April) to review these recommendations.

The EPA agrees with the intent of the recommendation regarding regional collaboration. However, the EPA is concerned that, as worded, it may not achieve the intended goal. We can accomplish increased regional collaboration through multiple avenues and, as such, recommend the following, more encompassing, language:

- The Administrator of EPA, in cooperation with other members of the tribal infrastructure task force, should identify and pursue additional mechanisms to increase collaboration at the regional level.

As the ITF lead agency, the EPA will work with our federal partners to identify and disseminate successful best practices at the regional level.

In conclusion, the EPA agrees with the GAO's analysis as it pertains to the EPA, and supports the recommendations to the Agency with the noted revision. The EPA will continue to find opportunities to improve water and wastewater infrastructure in Indian country and to collaborate with the ITF members at the headquarters and regional levels to ensure efficient and effective implementation of its tribal infrastructure programs. Thank you for the opportunity to review the draft report. For further technical details, questions, or additional information, please contact either Sam Russell (Russell.Sam@epa.gov; 202 564 4012) or Matthew Richardson (Richardson.Matthew@epa.gov; 202 564 2947).

Sincerely,

D Ross

David P. Ross
Assistant Administrator

cc: EPA GAO Liaison Team
Bobbie Trent
Mark Howard
Felicia Wright

APPENDIX VIII: COMMENTS FROM THE U.S. DEPARTMENT OF AGRICULTURE

USDA
United States Department of Agriculture

Rural Development

Office of the Assistant to the Secretary

1400 Independence Ave SW
Washington, DC 20250

Voice 202.720.4581
Fax 202.720.2080

April 2, 2018

Anne-Marie Fennell
Director
Natural Resources and Environment
U. S. Government Accountability Office

Dear Ms. Fennell:

Thank you for the opportunity to review and respond to Government Accountability Office (GAO) draft report GAO-18-309, "Drinking Water and Wastewater Infrastructure: Opportunities Exist to Enhance Federal Agency Needs Assessment and Coordination on Tribal Projects." USDA is pleased that GAO recognizes the efforts by USDA and other agencies to address water and waste water disposal infrastructure needs in tribal communities throughout Alaska and the continental United States. USDA's commitment and significant investments in tribal infrastructure are well documented. Over the last decade, we have made a concerted effort to improve processes and expand collaboration with our federal and state funding partners. It is important to USDA and the tribal communities that it serves, that this report fairly and accurately depict the delivery of USDA's critical infrastructure programs. Of particular concern is GAO's opinion that USDA is issuing grants to entities not authorized to directly receive them through the Rural Alaska Village Grant program. USDA disagrees with GAO's opinion and provides within this document the legal arguments to support its position.

USDA Rural Development (USDA) also submits the following comments, corrections, and suggested edits for consideration.

Native American Grants
GAO was asked to review how federal agencies identify tribal infrastructure needs, provide funding to address them, "including projects to address the most severe sanitation deficiencies," and collaborate to meet tribal infrastructure needs. GAO's report and recommendations focus primarily on the priority agencies have given to addressing severe deficiencies, regardless of whether the agency mission or congressional authorization exclusively requires it. Below are some requested corrections and edits that USDA believes more accurately reflect the nature of the USDA programs, while still providing a spotlight on the importance of addressing severe water and wastewater sanitation deficiencies.

1. USDA requests that the second sentence of paragraph two on page 11 be edited as stated below to accurately reflect Section 306C of the Consolidated

Farm and Rural Development Act of 1972 (Con Act) and USDA program delivery:

While the USDA administers funds for the Native American program at the national level, the program is delivered locally by USDA Rural Development State and Area Offices that conduct outreach to tribes, provide application assistance, process applications, and service grants. The Native American program provides grants for water and wastewater infrastructure and services to rural and low-income tribal communities. Tribal residents must face significant health risks as a result of not having access to, <u>or not being served by</u>, adequate and affordable water supply systems or waste disposal facilities.

GAO has omitted the statutory phrase underlined above which makes clear Congress' intent that the USDA program provide funds to projects addressing more than just Indian Health Service's Sanitary Deficiency System (SDS) Level 4 and 5 type projects. The language above regarding program delivery at USDA is important to properly characterize the program's working relationship with tribal customers.

2. Paragraph one on page 12, the draft report is missing information regarding USDA's Technical Assistance and Training grants and Solid Waste Management grants that are available to, and serve tribal and Alaska Native communities. We request that the following language be inserted at the end of paragraph 1:

   USDA also administers a Technical Assistance and Training (TAT) Grant program, as well as a Solid Waste Management Grant (SWM) program, both of which are available to, and subscribed to by tribal and Alaska native communities. From 2007 to 2017, USDA awarded $11.86 million in SWM and $20.34 million in TAT grants to tribes and non-profit entities to help tribes with technical assistance and training needs related to water, waste water, and solid waste management.

   Without this language, the report does not accurately reflect the totality of available USDA programs from which tribes may benefit.

3. Paragraph two on page 16, does not accurately characterize USDA's understanding of tribal needs. We request that the second sentence be modified to read:

   USDA officials explained that tribes identify needs through the applications they submit to the program. USDA also identifies tribal needs through outreach to tribes and coordination with other agencies, such as the Indian Health Service.

4. Paragraph three on page 20, the report should acknowledge that the statute does not require all programs to prioritize projects that address the most severe sanitation deficiencies. Section 306C of the Con Act clearly states that the grants are to be available to communities where a significant portion of residents do not have access to, <u>or are not served by</u>, adequate and affordable water supply systems or waste disposal facilities. While SDS levels are certainly taken into account when reviewing project applications, the statute does not reference the SDS deficiency definitions and does not

identify correction of severe sanitation deficiencies as a purpose of USDA's Native American program. GAO should acknowledge that USDA is operating well within the guidelines of the statute.

5. Paragraph one on page 22 does not accurately reflect the scope of the USDA TAT and SWM grant programs. As written, the report only references one grant for $130,000, severely understating USDA's commitment to, and investment in tribal communities. USDA requests that the text be modified to include the language below:

   From 2007 to 2017, USDA awarded $11.86 million in SWM and $20.34 million in TAT grants to tribes and non-profit entities to serve the technical assistance and training needs of tribes related to water, waste water and solid waste management.

6. Footnote 63 needs to be corrected to reflect that none of USDA's water and waste loan or grant programs are authorized to provide funding for operations and maintenance. In addition, paragraph two, sentence one, on page 22 should be edited as stated below to avoid misleading or confusing readers:

   Most federal programs reviewed are not authorized to, and did not, provide financial assistance for routine operations and maintenance of installed community or individual infrastructure.

7. Paragraph one on page 23, the third sentence should be edited for accuracy to read:

   Finally, under the applicable USDA regulation and policy, and pursuant to statutory requirements, the Native American's program objective is.....

8. In draft recommendation number 4, GAO recommends that The Assistant to the Secretary of Agriculture for Rural Development implement a scoring factor, similar to that used in USDA's Colonias grant program that awards points for proposed Native American program grant projects that address health risks from lack of access to safe drinking water and wastewater disposal.

   USDA would like clarification as to what form of scoring factor would be acceptable to GAO to address this recommendation. Making a regulation change, such as the one made for the Colonias grant program, could take 18 months or longer to execute. USDA would prefer to utilize the Administrator discretionary points in the existing regulation to implement the scoring factor and determine whether it has a material impact on funding through the Native American program. Such an approach could be implemented at the start of Fiscal Year 2019, or possibly sooner. As GAO is aware, USDA has funded all approved Native American projects seeking funding in recent years. Adding additional points to projects to address a lack of access to safe drinking water and waste water disposal could be impactful in years when funding requests exceed available funding, but not in years when it does not. A permanent change in the regulation could also have the unintended effect of discouraging tribes with eligible projects from seeking funding if their project does not meet the SDS 4 or 5 level. This could lead to a delay in investments in needed infrastructure and increase the deficiency level in tribal areas where the SDS Level is 3 or lower.

Utilizing the existing flexibility in the regulation, rather than permanently changing the regulation, would meet GAO's intent more quickly and avoid additional regulatory delay.

Rural Alaska Village Grants

USDA disagrees with the statements in the report regarding awards made in the Rural Alaska Village Grant Program (RAVG). As stated in the report, the legal position of USDA is that the authorization for the Rural Alaska Village Grant Program gives the State of Alaska control over the use of the funds provided. From 1997, USDA has worked directly with officials of the State of Alaska regarding the disposition of these funds, as required by law. This has led, with the concurrence of the State, to have some of the funds provided directly to parties other than the State, including Alaska Native Villages, under a Memorandum of Understanding (MOU). MOU partners include the Department of Agriculture, Rural Utilities Service (RUS); the Department of Health and Human Services, Indian Health Service (IHS); the State of Alaska, Department of Environmental Conservation (DEC); and, the Alaska Native Tribal Health Consortium (ANTHC). These are the same villages that have been specifically referenced in the statutory language of the appropriations for this program as far back as the 1998 fiscal appropriation with the inclusion of the following phrase: "shall be for water and waste disposal systems for rural and native villages in Alaska pursuant to section 306D of such Act." The State of Alaska's concurrence with project funding going directly to communities is not only provided through its signing of the 2011 MOU, but also through individual MOUs for each project where funding is direct to a community.

Starting in fiscal year 2011, after discussions with the State of Alaska, the MOU referenced above became the vehicle for the State of Alaska to inform USDA regarding the manner in which these funds would be administered. Under this MOU, as agreed to by the State of Alaska, ANTHC was provided some of these funds directly, as co-applicant with the native villages, to assist these villages with their water systems projects. This MOU was ultimately codified in regulations promulgated in 2015 (see 7 CFR 1784) through a transparent notice and comment process. Starting in fiscal year 2012, the appropriation committees adjusted the appropriation language for the RAVG program to reflect the approach taken in the MOU, and ultimately the regulations, by specifically referencing consortiums in a proviso. In sum, section 306D provides an earmarked grant to the State of Alaska and at no time in this process has the State of Alaska not had control over the manner in which the funds have been distributed by USDA.

Section 306D (7 USC 1926d) provides the State of Alaska with the grant for water systems in rural and native villages. USDA has respected this provision by working with the State of Alaska and helping the State direct a portion of these grants to parties who are helping the State carryout the purposes of the section. USDA believes that this mechanism is consistent with the purposes of section 306D and the appropriations of section 306D funds.

In addition, USDA offers the following comments and suggested edits and corrections.

1. According to USDA records, only two grants have been made directly to villages since the signing of the 2011 MOU. The last sentence of paragraph one, page 29, refers to grants back to the inception of the program and provides a skewed depiction of current activities. USDA requests than GAO note the activity since the inception of the MOU. In addition, the report states that USDA awarded 32 grants to ANTHC in 2011 before the statutory language in the appropriations bill was enacted. We are unclear as to why GAO believe 32 grants were made to ANTHC in FY 2011. Our records indicate that seven grants were made to ANTHC through the RAVG program in FY 2011. USDA has enclosed a list of all grants to ANTHC in the program's history as supporting documentation for your records. Further, USDA would like to emphasize that the grants were made pursuant to and after execution of the 2011 MOU signed by the State of Alaska, ANTHC and others. USDA requests that the paragraph be modified to correct the number of grants and give the proper context.

2. Paragraph three, page 29 should be expanded to provide the proper context in which USDA led the effort to develop the MOU. GAO's draft report, as written, does not reflect the information provided in meetings with GAO on the topic. USDA requests to modify the second sentence of the paragraph to read:

   For example, a 2010 USDA Office of Inspector General report found that the state was not properly documenting project costs and was transferring funds between projects. The result was that projects were often underfunded, delayed, or halted, and service was not being delivered to Alaskan village residents.

   As GAO acknowledges, the need for water and waste infrastructure in Alaska villages is acute. The MOU has enabled steady progress toward addressing these acute needs through a collaborative approach with funding partners.

3. The second sentence on page 30 needs to be corrected. The RD RAVG program manager stated in interviews that the agency awards grants directly to Alaska native villages that have the capacity to administer them.

4. Recommendations 5 and 6 are unnecessary given that the agency is operating within its authorities. USDA requests that GAO remove these recommendations from the report.

Collaboration

USDA offers the following technical corrections and edits related to the collaboration discussion in the draft report.

1. Paragraph two, sentence three, on page 32 should reference all of the stated purposes of the MOU, rather than just the one regarding leveraging of funds. By only referencing one of the purposes, GAO does not provide the reader with a clear understanding of the

holistic approach to serving tribes that the tribal infrastructure task force (ITF) is implementing.

2. Under the leadership section on page 32, paragraph four, GAO does not acknowledge the significant role USDA played in the development and adoption of the uniform preliminary engineering report. GAO should be aware of USDA's leadership role in this regard, as it has been highlighted in other GAO reports. USDA requests that the second sentence of the paragraph be modified to read:

   For example, in 2013 ***under USDA's leadership****, the members of the Tribal Infrastructure Task force agreed to adopt a uniform preliminary engineering report (PER) template, a key supporting document required by multiple agencies in their project application and evaluation processes.*

   We also suggest that the following language be added at the end of the paragraph: *Since the template was adopted, USDA has led an interagency effort with both federal and state water and waste funding partners to design an interactive, online version of the PER. USDA funded the development of the ePER online system, which launched in FY 2018 and is available for use by all federal and state-based water and waste funders, including the members of the ITF.*

   The ePER effort demonstrates significant collaboration amongst the ITF members and others, and facilitates better collaboration going forward.

3. Recommendation 9, page 41 – The 2011 report referred to is a working group report, not an ITF committee report. Given the length of time that has passed since the working group recommendations were made, USDA requests the report recommendation be edited to read as follows:

   The Assistant to the Secretary of Agriculture for Rural Development, in cooperation with other members of the tribal infrastructure task force, should review the 2011 working group report and determine appropriate actions to be taken.

4. Recommendation 13 – USDA requests that the recommendation be modified as follows:

   The Assistant to the Secretary of Agriculture for Rural Development, in cooperation with other members of the tribal infrastructure task force, should facilitate and promote the identification of additional mechanisms to increase collaboration.

Anne Hazlett
Assistant to the Secretary for Rural Development
United States Department of Agriculture

DEPARTMENT OF DEFENSE

DEPARTMENT OF THE ARMY
OFFICE OF THE ASSISTANT SECRETARY
CIVIL WORKS
108 ARMY PENTAGON
WASHINGTON DC 20310-0108

MAR 22 2018

Ms. Anne-Marie Fennell
Director, Natural Resources & Environment
U.S. Government Accountability Office
441 G. Street, NW
Washington, D.C. 20548

Dear Ms. Fennell:

This is the Department of Defense (DoD) response to the Government Accountability Office (GAO) Draft Report 18-309, "DRINKING WATER AND WASTEWATER INFRASTRUCTURE: Opportunities Exist to Enhance Federal Agency Needs Assessment and coordination on Tribal Projects," dated February 23, 2018 (GAO Code 101077).

The DoD acknowledges receipt of the subject draft report and has no comments. The DoD appreciates the opportunity to review the report and wishes to thank those involved for their candor, professionalism and collaborative engagement throughout the development of this draft report.

Sincerely,

R. D. James
Assistant Secretary of the Army
(Civil Works)

INDEX

A

B

E

F

G

H

I

L

M

N

O

P

Q

R

S

T

U

V

W

Z

Related Nova Publications

US Drinking Water Regulation: History and Politics, 1914-2015

Author: Michael Zarkin

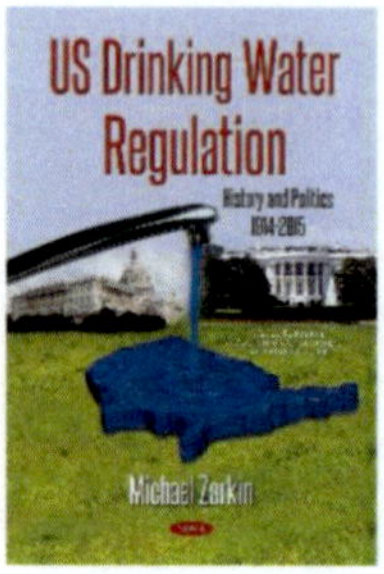

Series: Water Resource Planning, Development and Management

Book Description: In this book, Michael Zarkin traces the development of US drinking water standards, beginning with the earliest efforts by the US Public Health Service to craft national standards, and ending with the EPA's most recent efforts to implement the Safe Drinking Water Act.

Softcover ISBN: 978-1-53612-567-2
Retail Price: $82

Shared Water Resources of Lebanon

Authors: Amin Shaban and Mouin Hamzé

Series: Water Resource Planning, Development and Management

Book Description: This book aims to highlight the principles of Lebanon's water resources with new numeric measures. It also reveals the major elements of the striking challenges. Thus, the fundamental hydrologic aspects of shared water resources in Lebanon, including quantitative measures and the spatial extent of these resources are illustrated.

Hardcover ISBN: 978-1-53612-111-7
Retail Price: $95

To see a complete list of Nova publications, please visit our website at www.novapublishers.com